A-Level PHYSICS

Nuclear Physics and Fundamental Particles

Roger Muncaster

B.Sc Ph.D
Formerly Head of Physics
Bury Metropolitan College
of Further Education

Stanley Thornes (Publishers) Ltd.

First published in 1995 by
Stanley Thornes (Publishers) Ltd
Ellenborough House
Wellington Street
Cheltenham
Gloucestershire GL50 1YD
England

A catalogue record for this book is available from the British Library.

ISBN 07487 1805 2

The images on the front and back covers are as follows:

Front cover:

False-colour bubble chamber photograph showing annihilation of an antiproton (horizontal line entering at right) and a proton. The annihilation event has occurred at the intersection point just left of centre. The spiral tracks are due to low-energy electrons and positrons.

Back cover:

1. Computer simulation of an electron–positron collision in the DELPHI detector at CERN. The blue lines represent part of the cylindrical detector, and the variously coloured lines are the tracks of particles created in an electron–positron annihilation. The solid lines represent charged particles, which bend in the detector's magnetic field. The broken lines represent neutral particles, which are unaffected by the magnetic field.
2. False-colour bubble chamber photograph showing the associated production of two strange particles (see p. 96).
3. False-colour bubble chamber photograph showing the production and decay of a lambda particle. A high-energy proton (yellow) enters from the bottom, knocks out an atomic electron (green) in passing, then collides with a proton at rest in the liquid hydrogen of the bubble chamber. The collision produces 7 negative pions (blue), 9 positive particles (red), which include a proton and a kaon as well as 7 psi positive pions, and the lambda. Being neutral, the lambda leaves no track, but reveals its existence when it decays into a proton (yellow) and a negative pion (purple).

Typeset by Tech-Set, Gateshead, Tyne & Wear.
Printed and bound at Scotprint, Musselburgh

Contents

Preface

This book is intended to cover the various 'Nuclear Physics' and 'Particle Physics' options to be examined at A-level and AS-level from 1996 onwards. Specifically, it is aimed at Module Ph6 of the NEAB syllabuses in A-level and AS-level Physics, Module Ph4 (Topic 4c) of the London A-level Physics syllabus, Unit P8 of the Oxford and Cambridge syllabuses in A-level Physics and A-level Science, and Module 5 (Section B) of the Oxford syllabus in A-level Physics. Students following the NEAB End-of-Course scheme should also find the book useful, for it includes the material contained in that syllabus that is not currently found in general textbooks on A-level Physics.

The book contains many worked examples. Questions are included at relevant points in the text so that students can obtain an immediate test of their understanding of a topic. 'Consolidation' sections stress key points and in some cases present an overview of a topic in a manner which would not be possible in the main text. Definitions and fundamental points are highlighted – either by the use of screening or bold type. Questions, most of which are taken from past A-level papers, are included at the end of each of the nine chapters.

Acknowledgements

I wish to thank Veronica Hilton for her assistance with proof-reading and with the preparation of the index. I also wish to express my gratitude to the publishers, and to John Hepburn in particular, for their invaluable help throughout the preparation of the book. Many other people have helped me in a wide variety of ways – my sincere thanks to them.

I am indebted to the following examination boards for permission to use questions from their past examination papers:

Associated Examining Board [AEB]
University of Cambridge Local Examinations Syndicate [C], reproduced by permission of University of Cambridge Local Examinations Syndicate
Cambridge Local Examinations Syndicate, Overseas Examinations [C(O)]
Northern Examinations and Assessment Board (formerly the Joint Matriculation Board) [J]
Oxford and Cambridge Schools Examinations Board [O & C]
University of Oxford Delegacy of Local Examinations [O]
Southern Universities' Joint Board [S]
University of London Examinations and Assessment Council (formerly the University of London School Examinations Board) [L]
Welsh Joint Education Committee [W]

Thanks are also due to John Maple, of Joint European Torus, for his careful reading of the sections on nuclear fusion and for supplying the photograph on p. 77, and to the following organizations for providing photographs:

American Institute of Physics (Emilio Segrè Visual Archives): pp. 98, 100
Science Photo Library: pp. 3, 60, 93; back cover (3) (Brookhaven National Laboratory); pp. 123, 131 (CERN); pp. 88, 121 (David Parker); p. 107 (Hale Observatories); front cover, pp. 90, 96, back cover (2) (Lawrence Berkeley Laboratory); p. 43 (Novosti); back cover (1) (Philippe Plailly)

R. MUNCASTER
Helmshore

1
RADIOACTIVITY

1.1 STRUCTURE OF THE NUCLEUS, THE NUCLEONS

Every atom has a central, positively charged nucleus. Nuclear diameters are $\sim 10^{-15}$ m, atomic diameters are $\sim 10^{-10}$ m. Over 99.9% of the mass of an atom is in its nucleus. Atomic nuclei are unaffected by chemical reactions.

Nuclei contain **protons** and **neutrons**★ which, because they are the constituents of nuclei, are collectively referred to as **nucleons**. Their properties are compared with those of the electron in Table 1.1. The charge on the proton is equal and opposite to that on the electron, and it follows that (neutral) atoms contain equal numbers of protons and electrons.

Table 1.1
The nucleons compared with the electron

	Electron	Proton	Neutron
Mass	$m_e = 9.110 \times 10^{-31}$ kg	$m_p = 1836\, m_e$	$m_n = 1839\, m_e$
Charge	-1.602×10^{-19} C	$+1.602 \times 10^{-19}$ C	Zero

The nucleons are held together by one of the four fundamental forces – the **strong interaction**. (The other three are weaker than this, and in descending order of effectiveness are the electromagnetic force, the weak interaction and the gravitational force.) The strong interaction (sometimes called the **nuclear force**) is a very strong short-range force, and it more than offsets the considerable electrostatic repulsion of the positively charged protons.

1.2 ISOTOPES: DEFINITIONS

Two atoms which have the same number of protons but different numbers of neutrons are said to be **isotopes** of each other. It follows that each atom contains the same number of electrons as the other and, therefore, that their chemical properties are identical. Isotopes cannot be separated by chemical methods. Some elements have only one naturally occurring isotope (gold and cobalt are examples); tin has the largest number – ten.

> **The atomic number** (or **proton number**) Z of an element is the number of protons in the nucleus of an atom of the element.

The atomic number of an element was originally used to represent its position in the periodic table; it still does, but is now more meaningfully defined as above.

★The nucleus of the common isotope of hydrogen is an exception: it has a single proton and no neutron.

> **The mass number** (or **nucleon number**) A of an atom is the number of nucleons (i.e. protons + neutrons) in its nucleus.

The various isotopes of an element whose chemical symbol is represented by X are distinguished by using a symbol of the form

$$_Z^A \text{X}$$

where A and Z are respectively the mass number and atomic number of the isotope. The most abundant isotope of lithium (lithium 7) has 3 protons and 4 neutrons, i.e. $Z = 3$ and $A = 3 + 4 = 7$, and therefore it is represented by $_3^7\text{Li}$. Lithium 6 has only 3 neutrons and is written as $_3^6\text{Li}$. The Z-value is sometimes omitted because it gives the same information as the chemical symbol (for example, all lithium atoms have 3 protons).

Hydrogen is exceptional in that its three isotopes are given different names. The most abundant isotope has one proton and no neutron and is actually called hydrogen ($_1^1\text{H}$). The other isotopes are **deuterium** ($_1^2\text{D}$ or $_1^2\text{H}$) and **tritium** ($_1^3\text{T}$ or $_1^3\text{H}$). A deuterium nucleus is called a **deuteron**.

The relative atomic mass A_r of an atom is defined by

$$A_r = \frac{\text{Mass of atom}}{\text{One twelfth the mass of a } _6^{12}\text{C atom}} \qquad [1.1]$$

It follows that the relative atomic mass of $_6^{12}\text{C}$ is exactly twelve, that of $_1^1\text{H}$ is 1.008, that of $_8^{16}\text{O}$ is 15.995, etc. Note that **the relative atomic mass of an atom is approximately equal to the mass number of the atom**.

The unified atomic mass unit (u) is defined such that the mass of $_6^{12}\text{C}$ is 12 u (exactly). It follows that the mass of an atom expressed in unified atomic mass units is numerically equal to its relative atomic mass. The mass of a $_6^{12}\text{C}$ atom is found by experiment to be 1.993×10^{-26} kg, and therefore

$$1\text{u} = 1.661 \times 10^{-27} \text{ kg}$$

Note   The definition of relative atomic mass given above (equation [1.1]) applies to a single atom or single isotope. An alternative definition, which takes account of the natural isotopic composition of the element concerned, is

$$A_r = \frac{\text{Average mass of an atom of the element}}{\text{One twelfth the mass of a } _6^{12}\text{C atom}} \qquad [1.2]$$

When dealing with a single isotope (which is usually the case in nuclear physics) equation [1.1] should be used. Equation [1.2] is relevant in situations such as chemical reactions, where an element (normally) has its natural isotopic composition. Consider, for example, the case of uranium. On the basis of equation [1.1] the relative atomic mass of $_{92}^{235}\text{U} = 235.04$ and that of $_{92}^{238}\text{U} = 238.05$. The relative atomic mass of uranium, on the other hand, is given by equation [1.2] as 238.03. This is very close to the value for $_{92}^{238}\text{U}$, because over 99% of naturally occurring uranium is in the form of $_{92}^{238}\text{U}$.

The term **nuclide** is used to specify an atom with a particular number of protons and a particular number of neutrons. Thus ^6_3Li, ^7_3Li, $^{16}_8\text{O}$ and $^{18}_8\text{O}$ are four different nuclides.

Isotopes are nuclides with the same number of protons.

Isotones are nuclides with the same number of neutrons.

Isobars are nuclides with the same number of nucleons.

QUESTIONS 1A

1. How many of each of the following particles are there in a single atom of iron 56 ($^{56}_{26}\text{Fe}$)?
   (a) electrons,
   (b) protons,
   (c) neutrons,
   (d) nucleons,
   (e) negatively charged particles,
   (f) positively charged particles,
   (g) neutral particles.

2. The two naturally occurring isotopes of copper have relative atomic masses of 62.9296 and 64.9278. (a) What are the mass numbers of these isotopes? (b) What are their masses in unified atomic mass units? (c) Calculate the relative atomic mass of naturally occurring copper, given that the isotope with a relative atomic mass of 62.9296 is 69.0% abundant.

1.3 RADIOACTIVE DECAY

In 1896 Becquerel noticed that some photographic plates which had been stored close to a uranium compound had become fogged. He showed that the fogging was due to 'radiations'* emitted by the uranium. The phenomenon is called **radioactivity** or **radioactive decay** and the 'radiations' are emitted when an unstable nucleus disintegrates to acquire a more stable state. **The disintegration is spontaneous** and most commonly involves the emission of an **α-particle** or a **β-particle**. In both α-emission and β-emission the **parent nucleus** (i.e. the emitting nucleus) undergoes a change of atomic number and therefore becomes the nucleus of a different element. This new nucleus is called the **daughter nucleus** or the **decay product**. It often happens that the daughter nucleus is in an excited state when it is formed, in which case it reaches its ground state by emitting a third type of

Antoine Henri Becquerel
(1852–1908)

*Many of the 'radiations' are in fact particles; the term was applied before this was realized and is still in use.

radiation called a γ-**ray**. The emission of a γ-ray simply carries away the energy released when the daughter nucleus undergoes its transition to the ground state. Some radioactive substances decay in such a way that the daughter nuclei are produced in their ground states and therefore do not give any γ-emission. Though most nuclides emit either α-particles or β-particles, some emit both. For example 64% of $^{212}_{83}\text{Bi}$ nuclei emit β-particles and 36% emit α-particles.

1.4 α-PARTICLES (SYMBOL $^4_2\alpha$ OR $^4_2$He)

An α-particle consists of two protons and two neutrons, i.e. it is identical to a helium nucleus. The velocity with which an α-particle is emitted depends on the species of nucleus which has produced it and is typically 6% of the velocity of light. This corresponds to a kinetic energy of 6 MeV, and the α-particles are the most energetic form of 'radiation' produced by radioactive decay. Many α-emitters produce α-particles of one energy only. Others emit α-particles with a small number of nearly equal, discrete values of energy. For example, $^{212}_{83}\text{Bi}$ emits α-particles with energies of 6.086, 6.047, 5.765, 5.622, 5.603, and 5.481 MeV; those with 6.086 MeV and 6.047 MeV account for over 97% of the emission. α-particles are emitted by heavy nuclei.

Since α-particles are charged and move relatively slowly, they produce considerable ionization ($\sim 10^5$ ion-pairs* per cm in air at atmospheric pressure). As a consequence they lose their energy over a short distance and, for example, are capable of penetrating only a single piece of paper or about 5 cm of air.

When a nucleus undergoes α-decay it loses four nucleons, two of which are protons. Therefore

(i)   its mass number (A) decreases by 4, and

(ii)   its atomic number (Z) decreases by 2.

Thus, if a nucleus X becomes a nucleus Y as a result of α-decay, then

$$\underset{\text{(Parent)}}{^A_Z\text{X}} \longrightarrow \underset{\text{(Daughter)}}{^{A-4}_{Z-2}\text{Y}} + \underset{\text{(α-particle)}}{^4_2\alpha}$$

For example, uranium 238 decays by α-emission to thorium 234 according to

$$^{238}_{92}\text{U} \longrightarrow {}^{234}_{90}\text{Th} + {}^4_2\alpha$$

Note that the number of nucleons is conserved ($238 \rightarrow 234 + 4$) and the charge is conserved ($92 \rightarrow 90 + 2$).

The specific charge (i.e. the charge-to-mass ratio) of the α-particles was measured, soon after their discovery, by deflecting them in electric and magnetic fields. This showed that the particles were positively charged, and that their specific charge was the same as that of a doubly ionized helium atom (i.e. a helium nucleus). Confirmation that α-particles are helium nuclei was provided by Rutherford and Royds in 1909. Their apparatus is shown schematically in Fig. 1.1. Radon, an α-emitting gas, was contained in a thin-walled glass tube A. The walls of this tube were less than 0.01 mm in thickness and could be penetrated by the α-particles produced by the radon. The particles were incapable of passing through the thick outer wall B and so were trapped in C. Although C was evacuated, traces of air remained, and the

---

*Ionization results in the release of two charged particles – an electron and a positive ion, collectively known as an **ion-pair**.

Fig. 1.1
Rutherford and Royd's
apparatus to confirm
α-particles as helium
nuclei

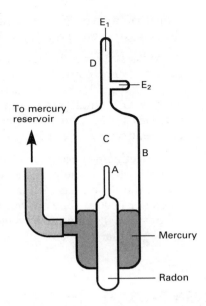

α-particles picked up electrons as a result of ionizing air molecules and by colliding with the walls. After about a week the gas which had accumulated in C was compressed and forced into the narrow tube D by raising the level of the mercury. An electrical discharge passed between the electrodes E_1 and E_2 caused the gas to produce its emission spectrum. When this was examined it was found to be the spectrum of helium – final proof that α-particles are helium nuclei.

1.5 β-PARTICLES (SYMBOL $_{-1}^{0}\beta$ OR $_{-1}^{0}e$)

These are very fast electrons (up to 98% of the velocity of light). In spite of their great velocities they have less energy than α-particles on account of their much smaller mass. **Any given species of nucleus emits β-particles with a continuous range of energies.** The maximum energies of the β-particles emitted by the naturally occurring nuclides vary from 0.05 MeV in the case of $_{88}^{228}$Ra to 3.26 MeV in the case of $_{83}^{214}$Bi.

β-particles are emitted by nuclei in which the neutron–proton ratio is too high for stability. Such a nucleus attains a more stable state (i.e. a lower energy state) when one of its neutrons changes into a proton and an electron. When this happens the electron is immediately emitted as a β-particle. The proton remains in the nucleus so that the nucleus has effectively lost a neutron and gained a proton, i.e. it has become a different element. There is no change in the total number of nucleons. Thus, when a nucleus undergoes β-decay:

(i)   its mass number (A) does not change, and

(ii)   its atomic number (Z) increases by 1, i.e.

$$_{Z}^{A}X \longrightarrow {}_{Z+1}^{A}Y + {}_{-1}^{0}e$$
$$\text{(Parent)} \qquad \text{(Daughter)} \qquad \text{(}\beta\text{-particle)}$$

For example, carbon 14 decays by β-emission to nitrogen 14 according to

$$_{6}^{14}C \longrightarrow {}_{7}^{14}N + {}_{-1}^{0}e$$

Note that, as in α-decay, the number of nucleons is conserved $(14 \to 14 + 0)$ and the charge is conserved $(6 \to 7 - 1)$.

β-particles are much less massive than α-particles and are much more easily deflected. The path of a β-particle through matter is therefore tortuous. Because they move quickly and are easily deflected, β-particles spend very little time in the vicinity of a single atom and therefore produce much less ionization than α-particles ($\sim 10^3$ ion-pairs per cm in air at atmospheric pressure). The most energetic have ranges which are about 100 times those of α-particles.

By deflecting β-particles in electric and magnetic fields Kaufman (1902) was able to establish that the particles were negatively charged and that their charge-to-mass ratio decreased with increasing velocity. A later series of measurements by Bucherer (1909) showed that the variation was in excellent agreement with the supposition that the particles were electrons whose mass was varying with velocity in the manner predicted by the special theory of relativity.

1.6 γ-RAYS (SYMBOL $^0_0\gamma$ OR γ)

γ-rays are electromagnetic radiation of very short wavelength. The wavelength of the radiation is characteristic of the nuclide which produces it. Many nuclides produce γ-rays of more than one wavelength; these wavelengths do not form a continuous spectrum, but are limited to a few discrete values. The wavelengths of the γ-rays produced by naturally occurring radioactive nuclides are typically in the range 10^{-10} m to 10^{-12} m, corresponding to energies of about 0.01 MeV to about 1 MeV. (The γ-rays produced by cosmic rays may have a wavelength of less than 10^{-15} m, i.e. an energy of more than about 10^3 MeV.) It is not uncommon for X-rays to have a wavelength of 10^{-11} m. These differ from γ-rays of the same wavelength only in the manner in which they are produced; **γ-rays are a result of nuclear processes, whereas X-rays originate outside the nucleus.**

In comparison with α-particles and β-particles γ-rays produce very little ionization and are very penetrating. γ-rays cannot be deflected by electric and magnetic fields.

The properties of the 'radiations' are summarized in Table 1.2.

Table 1.2   Summary of the properties of the 'radiations'

Property	α-particle	β-particle	γ-ray
Nature	Helium nucleus	Fast electron	Electromagnetic radiation
Charge	$+3.2 \times 10^{-19}$ C	-1.6×10^{-19} C	0
Rest mass	6.4×10^{-27} kg $= 4.0015$ u	9.1×10^{-31} kg $= 0.00055$ u	0
Velocity	$\sim 0.06\,c$	Up to $0.98\,c$	c
Energy	~ 6 MeV	~ 1 MeV	$hf \sim 0.01$ MeV
Number of ion-pairs per cm of air	$\sim 10^5$	$\sim 10^3$	~ 10
Path through matter	Straight	Tortuous	Straight
Ability to produce fluorescence	Yes (strong)	Yes	Yes (weak)
Ability to affect a photographic plate	Yes	Yes	Yes
Penetration	~ 5 cm of air	~ 500 cm of air ~ 0.1 cm of aluminium	~ 4 cm of lead reduces intensity to 10%

QUESTIONS 1B

1. The following questions refer to α-particles, β-particles and γ-rays. Which: **(a)** are emitted by parent nuclei, **(b)** are emitted by daughter nuclei, **(c)** cause the nuclei emitting them to change **(i)** mass number, **(ii)** atomic number, **(d)** are electrons, **(e)** are helium nuclei, **(f)** typically have the most energy, **(g)** typically have the least energy, **(h)** have a continuous range of energies, **(i)** are most easily absorbed, **(j)** produce the least ionization, **(k)** have the greatest specific charge (i.e. the greatest charge-to-mass ratio), **(l)** have tortuous paths through matter, **(m)** can be deflected by both electric and magnetic fields, **(n)** have a range of about 500 cm in air?

2. Write down the numerical values of the letters a to i in the following equations.

$$^{224}_{88}\mathrm{Ra} \longrightarrow {}^{a}_{b}\alpha + {}^{c}_{d}\mathrm{Rn}$$

$$^{e}_{83}\mathrm{Bi} \longrightarrow {}^{f}_{-1}\beta + {}^{210}_{g}\mathrm{Po} + {}^{h}_{i}\gamma$$

1.7 THE ABSORPTION OF α, β AND γ BY MATTER

Absorption of α-particles

Many α-emitters produce α-particles of one energy only. The α-particles are said to be **monoenergetic**, and they have very nearly equal ranges in any particular absorber (Fig. 1.2).

Fig. 1.2
Range of α-particles in an absorber

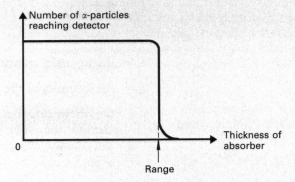

The range in air of the particles coming from a source can be found by determining the number* of α-particles that reach a detector when it is placed at different distances from the source. As the detector is moved away from the source the number of α-particles reaching it stays constant at first and then falls rapidly. An α-particle loses some of its energy each time it ionizes an air molecule. Since all the α-particles have the same energy initially, they each produce the same amount of ionization, but because of the randomness associated with the chance of an α-particle encountering an air molecule some travel a little further than others before giving up all their energy. This accounts for the tail on the curve and is known as **straggling**. The range is taken to be the maximum thickness that the majority of the α-particles can penetrate.

The range in aluminium (say) is found by placing successively thicker sheets of very thin aluminium foil between the source and the detector. (The source and detector are as close as possible so that there is no absorption due to air.) This also gives a curve of the type shown in Fig. 1.2. The range is reduced in aluminium, by approximately the ratio of the density of aluminium to that of air.

*In practice it is usually some quantity which is <u>proportional to the number</u> of particles that is measured.

Absorption of β-particles

The absorption of β-particles is complicated by the fact that the particles emitted by any nuclide have a <u>continuous</u> range of energies. Even if monoenergetic particles are selected by some means, it is still difficult to assess their range. This is because:

(i)   when β-particles are absorbed they sometimes eject high-speed electrons from the atoms of the absorber and these are easily confused with genuine β-particles, and

(ii) β-particles are easily deflected and therefore many are scattered away from the detector and therefore are not counted even though they have not been absorbed.

Over a limited thickness of absorber, however, the absorption of β-particles is approximately exponential.

Absorption of γ-rays

Many nuclides emit γ-rays of more than one wavelength. If γ-rays <u>of a single wavelength</u> are selected, their absorption is an exponential function of absorber thickness, i.e.

$$I = I_0\,e^{-\mu d} \qquad\qquad\qquad\qquad [1.3]$$

where

I = the intensity transmitted by a thickness d of absorber (Fig. 1.3)

I_0 = the intensity of the γ-rays incident on the absorber

μ = the **linear absorption coefficient** (or **attenuation coefficient**) of the absorber. The value of μ depends on the nature of the absorber and the wavelengths of the γ-rays. (Unit = m^{-1}.)

Fig. 1.3
γ-rays passing through an absorber

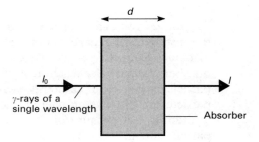

It follows from equation [1.3] that

$$\log_e I = \log_e (I_0\,e^{-\mu d})$$
$$= \log_e I_0 + \log_e (e^{-\mu d})$$

i.e. $\log_e I = -\mu d + \log_e I_0$

Plots of I against d, and $\log_e I$ against d are shown in Fig. 1.4. The gradient of the latter gives μ.

Fig. 1.4
(a) The effect of absorber thickness on the intensity of γ-rays transmitted.
(b) Plot to determine the linear absorption coefficient

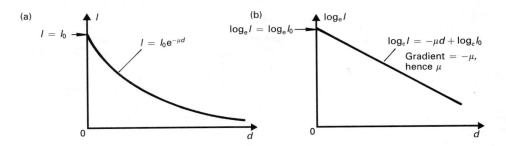

(a)
$I = I_0$
$I = I_0e^{-\mu d}$

(b)
$\log_e I = \log_e I_0$
$\log_e I = -\mu d + \log_e I_0$
Gradient $= -\mu$, hence μ

Notes

(i) The absorption of γ-rays increases with the atomic number of the material of the absorber. Atomic number has little effect on the absorption of α-particles and β-particles.

(ii) The exponential nature of γ-ray absorption arises because, in most cases, a γ-ray quantum loses all its energy in a single event, and therefore the fractional intensity of the beam falls by a fixed amount each time it traverses any given small thickness of absorber.

(iii) Equation [1.3] also holds for X-rays.

1.8 THE INVERSE SQUARE LAW FOR γ-RAYS

A point source of γ-rays emits in all directions about the source. It follows that the intensity of the γ-radiation decreases with distance from the source because the rays are spread over greater areas as the distance increases. This decrease in intensity is distinct from that produced by absorption.

Consider a point source of γ-rays, situated in a vacuum so that there is no absorption. The radiation spreads in all directions about the source, and therefore when it is a distance d from the source it is spread over the surface of a sphere of radius d and area $4\pi d^2$. If E is the energy radiated per unit time by the source, then the intensity of the radiation ($=$ energy per unit time per unit area) is given by I, where

$$I = E/(4\pi d^2) \qquad \text{i.e.} \quad I \propto 1/d^2$$

Thus the intensity varies as the inverse square of the distance from the source. The law is entirely true in vacuum. The absorption of γ-rays by air at atmospheric pressure is very slight, and therefore the inverse square law can be taken to hold over large distances in air.

Note The inverse square law holds for α-particles and β-particles <u>in vacuum</u> providing they are coming from a <u>point</u> source. The law does not apply to these particles in air because air absorbs them.

1.9 EXPERIMENTAL VERIFICATION OF THE INVERSE SQUARE LAW FOR γ-RAYS

The experimental arrangement is shown in Fig. 1.5. γ-rays may be absorbed at any point in the Geiger–Müller tube. Nevertheless, it is as if they are all absorbed at a single point (B). Neither the location of this point nor that of the source (A) are known and this makes it impossible to measure d directly. There is no real difficulty though, since, as can be seen from Fig. 1.5, $d = x + c$ and x is measurable and c, though unknown, is <u>constant</u>.

Fig. 1.5
Apparatus to verify the
inverse square law for
γ-rays

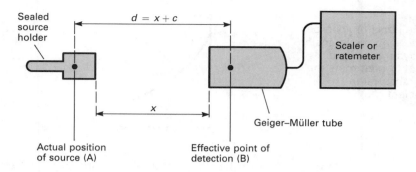

The aim is to verify that

$$I \propto \frac{1}{d^2}$$

i.e.

$$I \propto \frac{1}{(x+c)^2} \qquad\qquad [1.4]$$

Since I is proportional to the **corrected count rate** R (i.e. the actual count rate minus the background count rate), equation [1.4] can be rewritten as

$$R \propto \frac{1}{(x+c)^2}$$

i.e.

$$x + c = \frac{k}{R^{1/2}}$$

where k is a constant of proportionality

i.e.

$$x = kR^{-1/2} - c$$

If a plot of x against $R^{-1/2}$ turns out to be linear, the inverse square law has been verified. The intercept when $R^{-1/2}$ is zero gives c – see Fig. 1.6.

Fig. 1.6
Graph to verify the
inverse square law for
γ-rays

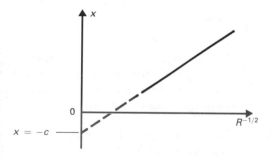

1.10 THE EXPONENTIAL LAW OF RADIOACTIVE DECAY

Radioactive nuclei disintegrate spontaneously; the process cannot be speeded up or slowed down. It follows that **for large numbers of any particular species of nuclei the rate of decay is proportional to the number of parent nuclei present**. If there are N parent nuclei present at time t, the rate of increase of N is dN/dt and therefore the rate of decrease, i.e. the rate of decay, is $-dN/dt$. It follows that

$$-\frac{dN}{dt} = \lambda N$$

or $\qquad \boxed{\dfrac{dN}{dt} = -\lambda N}$ [1.5]

where

λ = a (positive) constant of proportionality called **the decay constant**. It is equal to **the fraction of nuclei decaying per unit time**.$\star$ (Unit = s^{-1}.)

$-dN/dt$ = the rate of decay and is called the **activity** of the source. When used in equation [1.5] the activity must be expressed in the relevant SI unit – the becquerel. One **becquerel** (Bq) is equal to an activity of one disintegration per second. Activity used to be expressed in curies. One **curie** (Ci) is defined as (exactly) 3.7×10^{10} disintegrations per second.

Equation [1.5] can be rearranged as

(Note. The variables, N and t, are now on opposite sides of the equation.)

$$\frac{dN}{N} = -\lambda \, dt$$

Hence

$$\int \frac{dN}{N} = -\lambda \int dt$$

i.e. $\qquad \log_e N = -\lambda t + c$ [1.6]

If the initial number of nuclei is N_0, i.e. if $N = N_0$ when $t = 0$, then by equation [1.6]

$$\log_e N_0 = c$$

Substituting for c in equation [1.6] gives

$$\log_e N = -\lambda t + \log_e N_0$$

i.e. $\qquad \log_e N - \log_e N_0 = -\lambda t$

i.e. $\qquad \log_e \left(\dfrac{N}{N_0} \right) = -\lambda t$

i.e. $\qquad \dfrac{N}{N_0} = e^{-\lambda t}$

i.e. $\qquad \boxed{N = N_0 \, e^{-\lambda t}}$ [1.7]

Equation [1.7] expresses the exponential nature of radioactive decay, i.e. that the number of nuclei remaining after time t (i.e. the number of parent nuclei) decreases exponentially with time (see Fig. 1.7). It is known as the **exponential law of radioactive decay**.

$\star \, \lambda = \dfrac{-dN/dt}{N} = \dfrac{\text{Number of nuclei decaying per unit time}}{\text{Number of parent nuclei present}} = \text{Fraction decaying per unit time.}$

Fig. 1.7
Graph to illustrate the
exponential nature of
radioactive decay

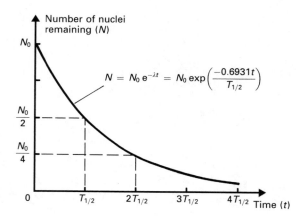

Thus, although it is impossible to predict when any particular nucleus will disintegrate, it is possible to say what proportion of a large number of nuclei will disintegrate in any given time.

1.11 HALF-LIFE ($T_{1/2}$)

If the life of a radioactive nuclide is taken to mean the time that elapses before all the nuclei present disintegrate, then it is clear from equation [1.7] (or from Fig. 1.7) that the life of any radioactive nuclide is infinite, i.e. $N = 0$ when $t = \infty$. It is not very useful, therefore, to talk about the life of a radioactive nuclide, and instead we refer to its half-life. **The half-life of a radioactive nuclide is the time taken for half the nuclei present to disintegrate.** If the half-life is represented by $T_{1/2}$, then when $t = T_{1/2}$, $N = N_0/2$, and therefore by equation [1.7]

$$\frac{N_0}{2} = N_0\, e^{-\lambda T_{1/2}}$$

i.e. $\frac{1}{2} = e^{-\lambda T_{1/2}}$

i.e. $\log_e \left(\frac{1}{2}\right) = -\lambda T_{1/2}$

i.e. $-0.6931 = -\lambda T_{1/2}$

i.e. $\boxed{T_{1/2} = \dfrac{0.6931}{\lambda}}$ [1.8]

The concept of half-life is illustrated in Fig. 1.7. The reader should verify, by selecting any point on the curve, that the number of nuclei halves whenever t increases by $T_{1/2}$. The half-life of any given nuclide is constant, in particular, it does not depend on the number of nuclei present. Half-lives have a very wide range of values; this can be confirmed by inspection of Table 1.3.

EXAMPLE 1.1

A sample of a radioactive material contains 10^{18} atoms. The half-life of the material is 2.000 days. Calculate:

(a)   the fraction remaining after 5.000 days,

(b)   the activity of the sample after 5.000 days.

Solution

(a)   Since $N = N_0\,e^{-\lambda t}$, the fraction, N/N_0, remaining after time t is given by

$$\frac{N}{N_0} = e^{-\lambda t}$$

Here $t = 5.000$ days and $\lambda = 0.6931/2.000$ day^{-1}.

$$\therefore \quad \lambda t = \frac{0.6931}{2.000} \times 5.000 = 1.7328$$

$$\therefore \quad \frac{N}{N_0} = e^{-1.7328} = 0.1768$$

i.e.   Fraction remaining after 5.000 days $= 0.1768$

(There has been no need to express t in s, nor to express λ in s^{-1}. We are concerned with λt, which is a pure number and therefore any unit of time can be used for t as long as the reciprocal of the same unit is used for λ.)

(b) $$\frac{dN}{dt} = -\lambda N$$

Here

$$N = 0.1768 \times 10^{18} \quad \text{and} \quad \lambda = \frac{0.6931}{2.000 \times 24 \times 3600}\,\text{s}^{-1}$$

$$\therefore \quad \frac{dN}{dt} = -\frac{0.6931 \times 0.1768 \times 10^{18}}{2.000 \times 24 \times 3600}$$

$$= -7.092 \times 10^{11}\,\text{s}^{-1}$$

i.e.   Activity after 5.000 days $= 7.092 \times 10^{11}$ Bq

EXAMPLE 1.2

An isotope of krypton $\left(^{87}_{36}\text{Kr}\right)$ has a half-life of 78 minutes. Calculate the activity of $10\,\mu\text{g}$ of $^{87}_{36}\text{Kr}$. (The Avogadro constant, $N_A = 6.0 \times 10^{23}\,\text{mol}^{-1}$.)

Solution

To a good approximation, the relative atomic mass of an atom is equal to the mass number of the atom. It therefore follows that

$$87\,\text{g of }^{87}_{36}\text{Kr contain } 6.0 \times 10^{23} \text{ atoms}$$

$$\therefore \quad 10\,\mu\text{g of }^{87}_{36}\text{Kr contain } \frac{10 \times 10^{-6} \times 6.0 \times 10^{23}}{87} \text{ atoms}$$

$$= 6.90 \times 10^{16} \text{ atoms}$$

$$\frac{dN}{dt} = -\lambda N$$

Here $N = 6.90 \times 10^{16} \quad \text{and} \quad \lambda = \frac{0.6931}{78 \times 60}\,\text{s}^{-1}$

$$\therefore \quad \frac{dN}{dt} = -\frac{0.6931}{78 \times 60} \times 6.90 \times 10^{16}$$

$$= -1.02 \times 10^{13}\,\text{s}^{-1}$$

i.e.   Activity $= 1.02 \times 10^{13}$ Bq

EXAMPLE 1.3

A sample of radioactive material has an activity of 9.00×10^{12} Bq. The material has a half-life of 80.0 s. How long will it take for the activity to fall to 2.00×10^{12} Bq?

Solution

Since activity is proportional to the number of parent nuclei present (see section 1.10), it follows from

$$N = N_0\, e^{-\lambda t}$$

that

$$A = A_0\, e^{-\lambda t}$$

where A = activity at time t, and A_0 = activity at $t = 0$. Rearranging gives

$$\frac{A}{A_0} = e^{-\lambda t}$$

$$\therefore \quad \log_e\left(\frac{A}{A_0}\right) = -\lambda t$$

$$\therefore \quad \log_e\left(\frac{2.00 \times 10^{12}}{9.00 \times 10^{12}}\right) = -\frac{0.6931}{80.0} \times t$$

$$\therefore \quad -1.504 = -8.664 \times 10^{-3}\, t$$

$$\therefore \quad t = 174\,\text{s}$$

i.e. Time for activity to fall to 2.00×10^{12} Bq is 174 s.

QUESTIONS 1C

1. The half-life of a particular radioactive material is 10 minutes. <u>Without using a calculator,</u> determine what fraction of a sample of the material will decay in 30 minutes.

2. A Geiger counter placed 20 cm away from a point source of γ-radiation registers a count rate of $6000\,\text{s}^{-1}$. What would the count rate be 1.0 m from the source?

3. A radioactive source has a half-life of 20 s, and an initial activity of 7.0×10^{12} Bq. Calculate its activity after 50 s have elapsed.

4. A sample of radioactive waste has a half-life of 80 years. How long will it take for its activity to fall to 20% of its current value?

5. Potassium 44 $\left(^{44}_{19}\text{K}\right)$ has a half-life of 20 minutes and decays to form $^{44}_{20}\text{Ca}$, a stable isotope of calcium.
   (a) How many atoms would there be in a 10 mg sample of potassium 44?
   (b) What would be the activity of the sample?
   (c) What would the activity be after one hour?
   (d) What would the ratio of potassium atoms to calcium atoms be after one hour?
   ($N_A = 6.0 \times 10^{23}\,\text{mol}^{-1}$.)

6. What mass of radium 227 would have an activity of 1.0×10^6 Bq. The half-life of radium 227 is 41 minutes. ($N_A = 6.0 \times 10^{23}\,\text{mol}^{-1}$.)

1.12 THE RADIOACTIVE SERIES

Most of the radioactive nuclides which occur naturally have atomic numbers which are greater than that of lead and each of these can be arranged in one or other of three series – **the radioactive series**. They are known as the uranium series, the thorium series, and the actinium series. The first eight members of the uranium series are listed in Table 1.3. Thus $^{238}_{92}$U decays by α-emission to produce $^{234}_{90}$Th which is itself unstable and undergoes β- decay to give $^{234}_{91}$Pa, etc. The final member of each series is a stable isotope of lead – a different isotope in each case.

Thorium series	$^{232}_{90}\text{Th} \rightarrow {}^{208}_{82}\text{Pb}$
Uranium series	$^{238}_{92}\text{U} \rightarrow {}^{206}_{82}\text{Pb}$
Actinium series	$^{235}_{92}\text{U} \rightarrow {}^{207}_{82}\text{Pb}$

The members of any one series are exclusive to that particular series. This is because whenever the mass number of a nuclide changes, it changes by 4 (α-emission) and none of the parents is in the decay sequence of either of the other two.

Table 1.3
The first eight members of the uranium series

Element	Nuclide	Half-life	Radiation		Energy of α or β in MeV
Uranium	$^{238}_{92}$U	4.51×10^9 years	α	γ	4.2
Thorium	$^{234}_{90}$Th	24.1 days	β	γ	0.19
Protactinium	$^{234}_{91}$Pa	6.75 hours	β	γ	2.3
Uranium	$^{234}_{92}$U	2.47×10^5 years	α	γ	4.77
Thorium	$^{230}_{90}$Th	8.0×10^4 years	α	γ	4.68
Radium	$^{226}_{88}$Ra	1620 years	α	γ	4.78
Radon	$^{222}_{86}$Rn	3.82 days	α		5.49
Polonium	$^{218}_{84}$Po	3.05 minutes	α		6.0

There is a fourth series – the **neptunium series** $\left({}^{237}_{93}\text{Np} \rightarrow {}^{209}_{83}\text{Bi} \right)$ but this does not occur naturally since the half-life of its longest lived member $\left({}^{237}_{93}\text{Np} \right)$ is short compared with the age of the Earth.

Of the elements whose atomic numbers are <u>less</u> than that of lead, indium and rhenium are the only ones whose most abundant isotopes are radioactive. The half-life is very long in each case – 6×10^{14} years for $^{145}_{49}$In and $>10^{10}$ years for $^{187}_{75}$Re.

1.13 RADIOACTIVE EQUILIBRIUM

Consider a freshly produced sample of uranium 238 (the parent of the uranium series). Uranium 238 decays to produce thorium 234. Initially the rate of production of thorium will exceed the rate at which it is decaying, and the thorium content of the sample will increase. As the amount of thorium increases, its activity increases and eventually a situation is reached in which the rate of production of thorium is equal to its rate of decay. The half-life of uranium 238 is very much greater than the half-lives of its decay products, and therefore to a good approximation the rate of production of thorium is constant. It follows, therefore, that once the rate of decay of the thorium has become equal to its rate of production, the quantity of thorium in the sample will remain constant.

Thorium decays to produce protactinium 234, and some time after the thorium content has become constant, the protactinium content will also stabilize. Eventually the amounts of each of the decay products in the series will be constant. The situation is known as radioactive equilibrium. At equilibrium the rate of decay of each nuclide is equal to its rate of production and it follows that <u>all</u> the rates of decay are equal. From equation [1.5] therefore

$$\lambda_1 N_1 = \lambda_2 N_2 = \ldots$$

where the subscripts $_1, _2, \ldots$ relate to the first, second, ... members of the series. Since $\lambda_1 = 0.6931/T_1$ and $\lambda_2 = 0.6931/T_2$, where T_1 and T_2 are the half-lives of the first and second members of the series, then

$$\frac{N_1}{T_1} = \frac{N_2}{T_2} = \ldots \tag{1.9}$$

1.14 ARTIFICIAL RADIOACTIVITY

Radioactive nuclides which do not occur in nature can be produced by bombarding naturally occurring nuclides with atomic particles – **notably with neutrons inside a nuclear reactor**. For example, neutron bombardment of the common isotope of beryllium $\left(^{9}_{4}\text{Be}\right)$ results in the creation of $^{9}_{3}\text{Li}$ – a radioactive isotope of lithium which does not occur in nature. The reaction is

$$^{9}_{4}\text{Be} + ^{1}_{0}\text{n} \longrightarrow ^{9}_{3}\text{Li} + ^{1}_{1}\text{p}$$

Other particles used in this way include protons, deuterons and α-particles.

Though many artificially produced radioactive nuclides decay in the same way as naturally occurring substances, i.e. by α-emission or (negative) β-emission, some decay by processes not found in nature, notably by emitting a (positive) β-particle (i.e. a **positron**) or by emitting a nucleon (see section 2.9).

Some artificial radioisotopes are produced simply to satisfy scientific curiosity, others are widely used in medicine and industry (see section 1.17).

1.15 SCHOOL LABORATORY SOURCES

The radioactive sources which are commonly used in schools are listed in Table 1.4. Activities of $5\,\mu\text{Ci}$ are typical.

Table 1.4
Sources used in schools

Element	Nuclide	Radiation	Comment
Americium	$^{241}_{95}\text{Am}$	α	Also emits γ-rays. These are of low energy and of no importance.
Plutonium	$^{239}_{94}\text{Pu}$	α	Also emits γ-rays. These are of low energy and of no importance.
Strontium	$^{90}_{38}\text{Sr}$	β	The β-particles actually come from yttrium – a decay product of strontium.
Cobalt	$^{60}_{27}\text{Co}$	γ	Also emits low energy β-particles. These are absorbed by the foil surrounding the source.
Radium	$^{226}_{88}\text{Ra}$	α β γ	The β-particles are actually produced by some of the decay products of the radium.

1.16 RADIOACTIVE DATING

Uranium Dating

The presence of $^{238}_{92}U$ in some rocks allows estimates of their ages to be made.

From the ratio of $^{206}_{82}Pb$ to $^{238}_{92}U$

$^{206}_{82}Pb$ is the stable end product of the uranium series (section 1.12). $^{238}_{92}U$ is the parent of the series and therefore for every uranium atom that has decayed since the rock containing the uranium was formed, one atom of $^{206}_{82}Pb$ will have been produced.★ The ratio of $^{206}_{82}Pb$ to $^{238}_{92}U$ can be used to determine the age of the rock.

As an example of the method suppose that in a particular sample of rock the ratio of $^{206}_{82}Pb$ to $^{238}_{92}U$ is 0.6. If

N_U = number of uranium atoms present and

N_{Pb} = number of lead atoms present, then

$N_U + N_{Pb}$ = number of uranium atoms present initially.

Therefore, from equation [1.7]

$$N_U = (N_U + N_{Pb}) e^{-\lambda t} \qquad [1.10]$$

where

λ = the decay constant of $^{238}_{92}U$

t = the time for which the uranium has been decaying.

Rearranging equation [1.10] gives

$$\frac{N_U + N_{Pb}}{N_U} = e^{\lambda t}$$

i.e. $1 + \dfrac{N_{Pb}}{N_U} = e^{\lambda t}$

But $N_{Pb}/N_U = 0.6$, and therefore

$$1.6 = e^{\lambda t}$$

in which case, from calculator

$$\lambda t = 0.4700$$

The half-life of $^{238}_{92}U$ is 4.5×10^9 years, and therefore

$$\lambda = \frac{0.6931}{4.5 \times 10^9}$$

i.e. $t = \dfrac{0.4700 \times 4.5 \times 10^9}{0.6931}$

i.e. $t = 3.1 \times 10^9$ years

If it is assumed that there was no lead present when the rock was formed, then the age of the rock is 3.1×10^9 years.

★This ignores the small number of uranium atoms which have disintegrated but which are still in the process of becoming lead.

From the ratio of helium to $^{238}_{92}U$

The decay process which eventually converts $^{238}_{92}U$ to $^{206}_{82}Pb$ involves the emission of eight α-particles and six β-particles. If it is assumed that the α-particles remain in the uranium-bearing rock as helium atoms, then the ratio of helium to uranium can be used to determine the age of the rock.

Carbon 14 Dating

The common isotope of carbon is the stable isotope $^{12}_{6}C$. A radioactive form of carbon, $^{14}_{6}C$, is formed in the upper atmosphere by neutron bombardment of $^{14}_{7}N$, the common isotope of nitrogen. The reaction is

$$^{14}_{7}N + {}^{1}_{0}n \rightarrow {}^{14}_{6}C + {}^{1}_{1}p$$

The neutrons are produced by the interaction of cosmic rays with atmospheric nuclei. If it is assumed that cosmic ray activity has been constant for a period which is long compared with the half-life of $^{14}_{6}C$ (5730 years), then for some considerable time the rate of decay of $^{14}_{6}C$ must have been equal to its rate of production. This means that the ratio of $^{14}_{6}C$ to $^{12}_{6}C$ in the atmosphere will also have been constant for a considerable time.

Living matter takes in carbon in the form of carbon dioxide from the atmosphere. When an organism dies it ceases to take in atmospheric carbon and its $^{14}_{6}C$ content starts to decrease as a result of radioactive decay. The $^{12}_{6}C$ content, on the other hand, stays constant and therefore from the moment of death the ratio of $^{14}_{6}C$ to $^{12}_{6}C$ decreases.

The radioactivity of carbon can be used to date archaeological samples. For example, the activity of a given mass of carbon taken from an ancient piece of wood can be compared with that of an equal mass of carbon from (say) a living plant. From this comparison it is possible to estimate the time that has passed since the wood was part of a living tree.

As an example of the method suppose that an archaeological sample has an activity of 7.5 disintegrations per minute, and that an equal mass of carbon from a living plant has an activity of 15 disintegrations per minute. The activity of the sample is one half that of the present-day level and therefore its age is equal to the half-life of $^{14}_{6}C$, i.e. the sample is 5730 years old.

Matter which has been dead for more than about 12 000 years cannot be dated in this way since its $^{14}_{6}C$ content is too low to allow an accurate analysis.

1.17 USES OF RADIOACTIVITY

(i)   Cancer cells can be destroyed by γ-radiation from a high-activity source of cobalt 60. Deep-lying tumours can be treated by planting radium 226 or caesium 137 inside the body close to the tumour.

(ii) The thickness of metal sheet can be monitored during manufacture by passing it between a γ-ray source and a suitable detector. The thicker the sheet the greater the absorption of γ-rays.

(iii) The exact position of an underground pipe can be located if a small quantity of radioactive liquid is added to the liquid being carried by the pipe. This also allows leaks to be detected; the soil close to the leak becomes radioactive. Provided a short-lived radioisotope is used there is no permanent contamination of the soil.

(iv) A radioisotope is <u>chemically</u> identical to a non-active isotope of the same element, and therefore takes part in the same chemical reactions. Thus the rate at which iodine passes through the thyroid can be determined by feeding radioactive iodine to a patient and externally monitoring the subsequent radioactivity of the thyroid. Radioactive phosphorus is used to assess the different abilities of plants to take up phosphorus from different types of phosphate fertilizer.

(v) Radioactive dating: see section 1.16.

1.18 RADIATION HAZARDS

The cells of the body may undergo dangerous physical and chemical changes as a result of exposure to radiation. The extent of the damage depends on:

(i)   the nature of the radiation,

(ii) the part of the body exposed to the radiation,

(iii) the dose received.

α-particles are absorbed in the dead surface layers of the skin and therefore do not constitute a serious hazard unless their source is taken into the body. γ-rays can penetrate deeply into the body and are a serious hazard.

The energy absorbed by unit mass of irradiated material is called the **radiation dose**. The unit is the **gray** (Gy) and $1\,\text{Gy} = 1\,\text{J}\,\text{kg}^{-1}$. Until recently the unit was the **rad**. ($1\,\text{Gy} = 100\,\text{rad}$.)

In order to take account of the different biological effects of the different radiations it is useful to define the **effective dose** as

$$\text{Effective dose} = \text{Radiation dose} \times \text{RBE}$$

where RBE is the **relative biological effectiveness** of the radiation concerned. Approximate values are listed in Table 1.5.

Table 1.5
RBE values of some radiations

Radiation	RBE
$\beta\ \gamma$ X	1
n (slow)	3
n (fast)	10
α	10–20
Fission product	20

CONSOLIDATION

Nuclei contain protons and neutrons.

A nucleon is a proton or a neutron.

The strong interaction (or **nuclear force**) is the force that holds nucleons together inside the nucleus.

Atoms which have the same number of protons are atoms of the same **element**.

Two atoms which have the same number of protons but different numbers of neutrons are **isotopes** of each other.

Atomic number (or **proton number**) Z of an element is the number of protons in the nucleus of an atom of the element.

Mass number (or **nucleon number**) A of an atom is the number of nucleons (i.e. protons + neutrons) in its nucleus.

Relative atomic mass (A_r)

$$A_r = \frac{\text{Mass of atom}}{\text{One twelfth the mass of a } _6^{12}\text{C atom}} \qquad [1.1]$$

or

$$A_r = \frac{\text{Average mass of an atom of the element}}{\text{One twelfth the mass of a } _6^{12}\text{C atom}} \qquad [1.2]$$

Equation [1.1] applies to a single atom or single isotope. Equation [1.2] applies when different isotopes of the element are present.

The relative atomic mass of an atom is approximately equal to the mass number of the atom.

The unified atomic mass unit (u) is one twelfth the mass of a $_6^{12}\text{C}$ atom.

The mass of an atom in unified atomic mass units is numerically equal to its relative atomic mass.

α-particles ($_2^4\alpha$ or $_2^4\text{He}$) are identical to helium nuclei.

β-particles ($_{-1}^{0}\beta$ or $_{-1}^{0}\text{e}$) are (fast) electrons.

γ-rays ($_0^0\gamma$ or γ) are very short wavelength electromagnetic radiation.

Properties of α, β and γ are summarized in Table 1.2.

α-emission   A new element is formed,

$$\begin{array}{ccc} _Z^A\text{X} & \longrightarrow & _{Z-2}^{A-4}\text{Y} & + & _2^4\text{He} \\ \text{(Parent)} & & \text{(Daughter)} & & (\alpha\text{-particle}) \end{array}$$

β-emission   A new element is formed,

$$\begin{array}{ccc} _Z^A\text{X} & \longrightarrow & _{Z+1}^{A}\text{Y} & + & _{-1}^{0}\text{e} \\ \text{(Parent)} & & \text{(Daughter)} & & (\beta\text{-particle}) \end{array}$$

γ-emission γ-rays are emitted by the daughter nucleus – no new element is formed.

α-particles and γ-rays are emitted with a single energy, or with a small number of discrete values of energy.

β-particles are emitted with a continuous range of energies.

Radioactive decay is spontaneous and therefore the rate of decay is proportional to the number of parent nuclei present. This leads to

$$\frac{\mathrm{d}N}{\mathrm{d}t} = -\lambda N$$

where λ is a constant of proportionality called the **decay constant**.

Solving this equation gives **the exponential law of radioactive decay**

$$N = N_0 e^{-\lambda t}$$

Because activity, A, and count-rate, R, are proportional to N, it follows that

$$A = A_0 e^{-\lambda t} \quad \text{and} \quad R = R_0 e^{-\lambda t}$$

Activity is the rate of decay, i.e. the number of disintegrations per second. The unit of activity is the becquerel.

One becquerel (Bq) $= 1$ disintegration per second.

$$\text{Activity} = -\frac{dN}{dt} = \lambda N$$

Half-life $(T_{1/2})$ is the time taken for half the nuclei present to disintegrate.

$$T_{1/2} = \frac{\log_e 2}{\lambda} = \frac{0.6931}{\lambda}$$

Inverse square law for γ-rays

The intensity, I, at a distance, d, from a point source of γ-rays is given by

$$I \propto \frac{1}{d^2} \quad \text{i.e.} \quad I = \frac{\text{constant}}{d^2}$$

The law is entirely true in vacuum, and is a good approximation in air.

QUESTIONS ON CHAPTER 1

1. Part of the actinium radioactive series can be represented as follows:

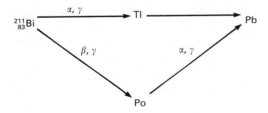

The symbols above the arrows indicate the modes of decay.
  (a) Write down the atomic numbers and mass numbers of Tl, Po and Pb in this series.
  (b) What is a possible mode of decay for the stage Tl to Pb? [C(O)]

2. Explain the following observations:
  (a) A radioactive source is placed in front of a detector which can detect all forms of radioactive emission. It is found that the activity registered is noticeably reduced when a thin sheet of paper is placed between the source and the detector.

  (b) A brass plate with a narrow vertical slit is now placed in front of the radioactive source and a horizontal magnetic field, normal to the line joining source and detector, is applied. It is found that the activity recorded is further reduced.
  (c) The magnetic field in (b) is removed and a sheet of aluminium is placed in front of the source. The activity recorded is similarly reduced.
  (d) The aluminium sheet in (c) is replaced by a sheet of lead and the detector records much less activity. This activity is not affected by the reintroduction of the magnetic field.
  [AEB, '83]

3. A certain α-particle track in a cloud chamber has a length of 37 mm. Given that the average energy required to produce an ion-pair in air is 5.2×10^{-18} J and that α-particles in air produce on average 5.0×10^3 such pairs per mm of track, find the initial energy of the α-particle. Express your answer in MeV.

(Electron charge $= 1.6 \times 10^{-19}$ C.) [C]

4. A sample of iodine contains 1 atom of the radioactive isotope iodine 131 (^{131}I) for every 5×10^7 atoms of the stable isotope iodine 127. Iodine has a proton number of 52 and the radioactive isotope decays into xenon 131 (^{131}Xe) with the emission of a single negatively charged particle.

(a) State the similarities and differences in composition of the nuclei of the two isotopes of iodine.

(b) What particle is emitted when iodine 131 decays? Write the nuclear equation which represents the decay.

(c) The diagram shows how the activity of a freshly prepared sample of the iodine varies

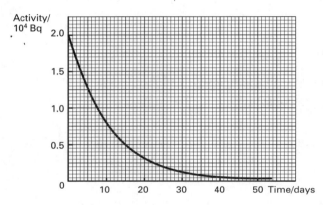

with time. Use the graph to determine the decay constant of iodine 131. Give your answer in s^{-1}.

(d) Determine the number of iodine atoms in the original sample.   [AEB, '91]

5. A radioactive source has a half-life of 20 days. Calculate the activity of the source after 70 days have elapsed if its initial activity is 10^{10} Bq.

6. The radioactive isotope $^{218}_{84}$Po has a half-life of 3 min, emitting α-particles according to the equation

$$^{218}_{84}\text{Po} \rightarrow \alpha + {}^{x}_{y}\text{Pb}.$$

What are the values of x and y?

If N atoms of $^{218}_{84}$Po emit α-particles at the rate of 5.12×10^4 s^{-1}, what will be the rate of emission after $\frac{1}{2}$ hour?   [S]

7. (a) In the uranium decay series $^{238}_{92}$U decays by stages to $^{234}_{92}$U.

(i) Why is the new nucleus given the symbol U?

(ii) Identify the total number and types of particles that have been emitted in this transformation.

(b) An isotope of the element radon has a half-life of 4 days. A sample of radon originally contains 10^{10} atoms.

(Take 1 day to be 86×10^3 s.)

Calculate:
(i) the number of radon atoms remaining after 16 days;
(ii) the radioactive decay constant for radon;
(iii) the rate of decay of the radon sample after 16 days.   [O, '92]

8. The half-life of $^{30}_{15}$P is 2.5 minutes. Calculate the mass of $^{30}_{15}$P which has an activity of 10^{15} Bq.

(The Avogadro constant $= 6.0 \times 10^{23}$ mol^{-1}.)

9. The activity of a particular radioactive nuclide falls from 1.0×10^{11} Bq to 2.0×10^{10} Bq in 10 hours. Calculate the half-life of the nuclide.

10. Calculate the activity of $2.0 \ \mu$g of $^{64}_{29}$Cu.

(The half-life of $^{64}_{29}$Cu $= 13$ hours. The Avogadro constant $= 6.0 \times 10^{23}$ mol^{-1}.)

11. The radioactive isotope of iodine, ^{131}I, has a half-life of 8.0 days and is used as a tracer in medicine. Calculate:

(a) the number of atoms of ^{131}I which must be present in a patient when she is tested to give a disintegration rate of 6.0×10^5 s^{-1},

(b) the number of atoms of ^{131}I which must have been present in a dose prepared 24 hours before.   [J, '91]

12 In 420 days, the activity of a sample of polonium, Po, fell to one-eighth of its initial value. Calculate the half-life of polonium. Give the numerical values of a, b, c, d, e, f in the nuclear equation.

$$^{a}_{b}\text{Po} \rightarrow {}^{c}_{d}\alpha + {}^{206}_{82}\text{Pb} + {}^{e}_{f}\gamma$$

13. A point source of γ-radiation has a half-life of 30 minutes. The initial count rate, recorded by a Geiger counter placed 2.0 m from the source, is 360 s^{-1}. The distance between the counter and the source is altered. After 1.5 hour the count rate recorded is 5 s^{-1}. What is the new distance between the counter and the source?   [L]

14. (a) (i) What are alpha, beta and gamma rays?
   (ii) Describe briefly one method whereby they may be distinguished from one another experimentally.

(b) Explain what is meant by:
   (i) radioactive decay,
   (ii) radioactive decay constant,
   (iii) half-life,
   (iv) the becquerel.

(c) (i) A newspaper article stated that the NASA Galileo space probe to Jupiter 'contained 49 lb of plutonium to provide 285 watts of electricity through its radioactive thermonuclear generator (RTG)'.

(Note: An RTG is a device for converting thermal energy produced by fission into electrical energy.)

Assuming that the plutonium is ^{239}Pu, which is built into a small nuclear reactor and that the efficiency of the RTG is 10%, what is the maximum time for which the RTG will supply the required energy output?
(Take the energy emitted for each nuclear disintegration of the ^{239}Pu to be 32 pJ, $N_A = 6.0 \times 10^{23}$ mol^{-1}, 1 lb = 0.45 kg.)

(ii) What factors will tend to: (I) increase, (II) decrease your estimate of the time?
                                                            [W, '91]

15. $^{210}_{83}$Bi is a *radioactive isotope* of bismuth with a *half-life* period of 5.00 days, which emits *negative beta particles*. Explain the italicized terms in the above statement and the significance of the numbers 210 and 83.

Describe experiments you would do to verify that alpha and gamma radiations from $^{210}_{83}$Bi are negligible. (You are not required to explain the principles underlying the action of any detector you may use.)

This isotope appears to be an ideal source of beta particles for an experiment you wish to perform. Assuming that for your experiments, which run continuously for 300 hours, the strength of the source must not fall below 10 μCi, what strength of source is required at the start of an experiment?   [O & C]

16. A decay sequence for a radioactive atom of radon-219 to a stable lead-207 atom is as shown below.

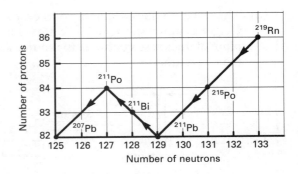

(a) What do the numbers on the symbol $^{207}_{82}$Pb represent?

(b) (i) Write down a nuclear equation representing the decay of $^{219}_{86}$Rn to $^{215}_{84}$Po.
   (ii) Write down the name of the particle which is emitted in this decay.

(c) (i) What particle is emitted when $^{211}_{83}$Bi decays?
   (ii) What happens within the nucleus to cause this decay?

(d) The half-life of $^{219}_{86}$Rn is 4.0 s. At time $t = 20$ s, what fraction of the radon atoms present at time $t = 0$, will be undecayed?   [C, '92]

17. The potassium isotope $^{42}_{19}$K has a half-life of 12 h, and disintegrates with the emission of a γ-ray to form the calcium isotope $^{42}_{20}$Ca. What other radiation besides γ-rays must be emitted? How many electrons, protons, and neutrons are there in an atom of the calcium isotope?
The amount of radiation received in unit time by a person working near a radioactive source, commonly called the dose rate, is measured in rem h^{-1}. The safety regulations forbid dose rates in excess of 7.5×10^{-4} rem h^{-1}. The γ-ray dose rate from the $^{42}_{19}$K source is found to be 3×10^{-3} rem h^{-1} at a distance of 1 m. What is the minimum distance from this source at which it is safe to work?
After how long will it be safe to work at a distance of 1 m from this source?   [S]

18. The isotope of strontium, $^{90}_{38}$Sr, decays by β^--emission and forms the radioactive isotope of yttrium, $^{90}_{39}$Y.
   (a) Write down an equation which describes the decay process.
   (b) The half-life of $^{90}_{38}$Sr is 17.8 years and the half-life of $^{90}_{39}$Y is 64.0 hours.
   Calculate the mass of yttrium present in a stable specimen (where the rates of decay of $^{90}_{38}$Sr and $^{90}_{39}$Y are equal) which contains 1.12 μg of $^{90}_{38}$Sr.

(c) The diagram shows a beta source near a Geiger counter and with a sheet of aluminium of thickness x between the source and the G–M tube.

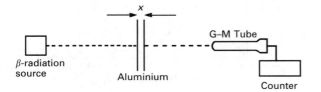

If I_o and I are the number of beta particles counted per second with the aluminium sheet absent and in place respectively, then it is expected that

$$I = I_o e^{-\mu x}$$

where μ is constant.

Given a supply of aluminium sheets of different known thicknesses, explain how you would test the above equation. How would you determine a value for μ?

[L, '93]

19. One of the decay products of radium, polonium 218, is an alpha emitter which decays as follows:

$$^{218}_{84}\text{Po} \rightarrow {}^{214}_{82}\text{Pb} + {}^{4}_{2}\text{He}$$

The alpha particles from the polonium can be used to convert aluminium 27 into a radioactive isotope of phosphorus, phosphorus 30, which decays into silicon 30.

$$^{27}_{13}\text{Al} + {}^{4}_{2}\text{He} \rightarrow {}^{30}_{15}\text{P} + \text{X}$$
$$^{30}_{15}\text{P} \rightarrow {}^{30}_{14}\text{Si} + \text{Y}$$

With reference to these three examples identify the particles X and Y, and explain the difference between natural and induced radioactivity.

The fundamental law of radioactive decay may be written

$$\frac{\mathrm{d}N}{\mathrm{d}t} = -\lambda N$$

State this equation in words.
The decay constant, λ, for a given material is related to its half-life, $T_{1/2}$, by the equation

$$\lambda = \frac{0.693}{T_{1/2}}$$

Define half-life.
The half-life of phosphorus-30 is 153 s. The activity of an aluminium disc after irradiation by

alpha particles is found to be 2500 Bq (where 1 Bq means one disintegration per second). Calculate the number of undecayed atoms of phosphorus-30 present in the aluminium disc.

[L, '93]

20. The decay of bismuth-209 is described by the equation:

$$^{209}_{83}\text{Bi} \rightarrow {}^{205}_{81}\text{Tl} + {}^{4}_{2}\text{He} + Q$$

In what form is the energy Q released?
Describe in general terms how
(a) momentum,
(b) charge, and
(c) mass/energy
are conserved during the decay process. (You may assume that the bismuth atom is stationary at the instant of decay.)

[L, '93]

21. The following account is taken from a science fiction magazine.

It is the distant future, and Man has long since abandoned Planet Earth. Space explorers land on the Earth's surface and discover a nuclear warhead, manufactured in the 20th century. They analyse a small sample of the warhead material and find it contains a mixture of radioactive plutonium Pu-239 and (effectively) stable uranium U-235.

The results for the sample are:

Mass of $^{239}_{94}\text{Pu}$ present $= 2.0 \times 10^{-6}$ kg

Mass of $^{235}_{92}\text{U}$ present $= 6.0 \times 10^{-6}$ kg

Activity of $^{239}_{94}\text{Pu}$ in sample $= 4.4 \times 10^{6}$ disintegrations s^{-1}.

(a) Calculate:
(i) the number of $^{239}_{94}\text{Pu}$ atoms in the sample.
(Take the Avogadro constant to be 6.0×10^{23} mol^{-1}, Molar mass of Pu-239 $= 0.239$ kg mol^{-1}.)
(ii) the radioactive decay constant, λ, for $^{239}_{94}\text{Pu}$.

(b) The explorers assume, correctly, that the material was originally pure plutonium, which decayed to produce the uranium.
(i) Approximately, what fraction of the original plutonium atoms remains undecayed?
(ii) The half-life of Pu-239 is 25 000 years. How far into the future is this account set?

[O, '93]

2

NUCLEAR STABILITY

2.1 THE ELECTRONVOLT

The electronvolt (eV) is a unit of energy. It is equal to the kinetic energy gained by an electron in being accelerated by a potential difference of one volt.

The work done when a particle of charge Q moves through a PD V is QV. The charge on the electron is 1.6×10^{-19} C, and therefore when an electron is accelerated through a PD of 1 V the work done is

$$(1.6 \times 10^{-19}) \times (1)$$

i.e. 1.6×10^{-19} J

The work done is equal to the kinetic energy gained by the electron, and therefore the kinetic energy gained by an electron in being accelerated through one volt is 1.6×10^{-19} J,

i.e.

$$1 \text{ eV} = 1.6 \times 10^{-19} \text{ J}$$

EXAMPLE 2.1

Write down the kinetic energy, in eV, of (a) an electron accelerated from rest through a PD of 10 V, (b) a proton accelerated from rest through a PD of 20 V, (c) a doubly charged calcium ion accelerated from rest through a PD of 30 V.

Solution

(a) 10 eV   (Since an electron accelerated through 1 V would gain 1 eV, it follows from $W = QV$ that an electron accelerated through 10 V gains 10 eV.)

(b) 20 eV   (Since the charge on the proton has the same magnitude as the charge on the electron, and an electron accelerated through 20 V would gain 20 eV.)

(c) 60 eV   (Since the charge on the calcium ion is twice that on an electron, and an electron accelerated through 30 V would gain 30 eV.)

QUESTIONS 2A

1. An electron is accelerated from rest through a PD of 1000 V. What is **(a)** its kinetic energy in eV, **(b)** its kinetic energy in joules, **(c)** its speed?
   ($1\text{eV} = 1.602 \times 10^{-19}$ J, mass of electron = 9.110×10^{-31} kg.)

2. An electron accelerated from rest through a PD of 50 V acquires a speed of $4.2 \times 10^6 \text{m s}^{-1}$. Without performing a detailed calculation, write down the speed that a PD of 200 V would produce.

3. What is the kinetic energy, in eV, of a triply charged ion of iron (Fe^{3+}) which has been accelerated from rest through a PD of 100 V?

4. An electron moves between a pair of electrodes in a vacuum tube. The first electrode is at a potential of 50 V, and the electron has kinetic energy of 20 eV as it leaves it. What is the potential of the second electrode if the electron just reaches it?

5. The kinetic energy of an α-particle from a radioactive source is 4.0 MeV. What is its speed?
   (Charge on electron = 1.6×10^{-19} C, mass of α-particle = 6.4×10^{-27} kg.)

2.2 EINSTEIN'S MASS–ENERGY RELATION

According to the special theory of relativity a mass m is equivalent to an amount of energy E, where

$$E = mc^2 \qquad\qquad [2.1]$$

c being the speed of light ($\approx 3 \times 10^8 \text{m s}^{-1}$).

It follows that whenever a reaction results in a release of energy there is an associated decrease in mass. For example, when 1 kg of $^{235}_{92}$U undergoes fission (see section 2.10) the energy released is approximately 8×10^{13} J, and therefore according to equation [2.1] there is a decrease in mass of $8 \times 10^{13}/(3 \times 10^8)^2 \approx 9 \times 10^{-4}$ kg. This is a significant fraction of the initial mass of $^{235}_{92}$U and can be measured. Chemical reactions, on the other hand, release relatively small amounts of energy and the associated decrease in mass is too small to be measured. For example, when 1 kg of petrol is burned the energy released is only 5×10^7 J and, by equation [2.1], this corresponds to a decrease in mass of a mere 5.5×10^{-10} kg.

The reader should be left in no doubt that no matter how a change in energy arises there is a change in mass. For example, an increase in temperature is accompanied by an increase in mass, as is an increase in velocity.

The unified atomic mass unit (u) is defined in section 1.2, and

$$1\,\text{u} \equiv 1.661 \times 10^{-27} \text{ kg}$$

Therefore, from equation [1.1]

$$1\,\text{u} = 1.661 \times 10^{-27} \times (2.998 \times 10^8)^2 \,\text{J}$$

$$= \frac{1.661 \times 10^{-27} \times (2.998 \times 10^8)^2}{1.602 \times 10^{-19}} \text{ eV} \approx 932 \,\text{MeV}$$

Using more accurate values of the various constants gives

$$1\,\text{u} = 931.5 \,\text{MeV} \qquad\qquad [2.2]$$

2.3 BINDING ENERGY

The mass of a nucleus is always less than the total mass of its constituent nucleons. The difference in mass is called the **mass defect** of the nucleus, i.e.

Mass defect = Mass of nucleons − Mass of nucleus

The reduction in mass arises because the act of combining the nucleons to form the nucleus causes some of their mass to be released as energy (in the form of γ-rays). Any attempt to separate the nucleons would involve them being given this same amount of energy – it is therefore called the **binding energy** of the nucleus. It follows from equation [2.1] that

$$\begin{array}{ccc} \text{Binding energy} = \text{Mass defect} \times & c^2 \\ \text{(J)} & \text{(kg)} & (\text{m s}^{-1})^2 \end{array}$$

It follows from equation [2.2] that

$$\begin{array}{cc} \text{Binding energy} = 931.5 \times \text{Mass defect} \\ \text{(MeV)} & \text{(u)} \end{array}$$

Tables normally give <u>atomic</u> masses rather than <u>nuclear</u> masses and it is useful to redefine the mass defect as

$$\begin{array}{c} \text{Mass} \\ \text{defect} \end{array} = \left(\begin{array}{c} \text{Mass of nucleons} \\ \text{and electrons} \end{array} \right) - \left(\begin{array}{c} \text{Mass of} \\ \text{atom} \end{array} \right)$$

Consider, as an example of the calculation of binding energies, the case of the helium atom. It consists of two protons (each of mass 1.007 28 u), two neutrons (each of mass 1.008 67 u) and two electrons (each of mass 0.000 55 u). The total mass of the particles is

$$2 \times 1.007\,28 + 2 \times 1.008\,67 + 2 \times 0.000\,55 = 4.033\,00\,\text{u}$$

The mass of a helium atom is 4.002 60 u, and therefore the mass defect is

$$4.033\,00 - 4.002\,60 = 0.030\,4\,\text{u}$$

From equation [2.2], therefore, the binding energy of a helium atom is

$$0.0304 \times 931.5 = 28.3\,\text{MeV}$$

Note   The binding energy of a <u>nucleus</u> is the energy required to break it up into its component neutrons and protons. The binding energy of an <u>atom</u>, on the other hand, is the energy required to break it up into its component neutrons, protons and electrons.

The difference between the two is negligible, because the energy required to remove the electrons is very much less than that required to remove the neutrons and protons. For example, the binding energy of a helium <u>nucleus</u> is also 28.3 MeV.

A useful measure of the stability of a nucleus is its **binding energy per nucleon** (i.e. binding energy divided by mass number), since this represents the (average) energy which needs to be supplied to remove a nucleon. Fig. 2.1 shows the way this quantity varies with mass number for the naturally occurring nuclides with mass numbers in the range 2–238. It can be seen that the nuclides of intermediate mass numbers have the largest values of binding energy per nucleon. $^{56}_{26}\text{Fe}$ has a value of

Fig. 2.1
Variation of binding
energy per nucleon with
mass number

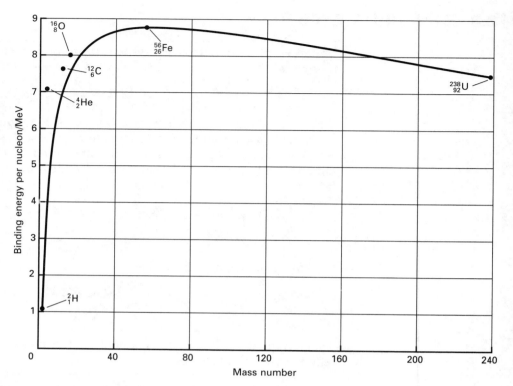

8.8 MeV and is one of the most stable nuclides. Three nuclides, $^4_2$He, $^{12}_6$C and $^{16}_8$O, lie significantly above the main curve. Though these are not the most stable of all nuclides, they are considerably more stable than those of adjacent mass. Note that $^{12}_6$C and $^{16}_8$O are respectively combinations of three and four α-particles. A combination of two α-particles would be $^8_4$Be. The binding energy per nucleon of $^8_4$Be is slightly less than that of an α-particle, and therefore it is unstable and disintegrates to form two α-particles.

Nuclides of intermediate mass number have the greatest binding energy per nucleon and therefore energy is released when two light nuclides are fused to produce a heavier one (**fusion**) and when a heavy nucleus splits into two lighter ones (**fission**).

QUESTIONS 2B

1. Calculate:
    (a) the mass defect,
    (b) the binding energy per nucleon for $^{238}_{92}$U.

    (Atomic mass of $^{238}_{92}$U $= 238.050\,76$ u,
    mass of neutron $= 1.008\,67$ u,
    mass of proton $= 1.007\,28$ u,
    mass of electron $= 0.000\,55$ u,
    1 u $= 931.5$ MeV.)

2.4 STABILITY AGAINST α-PARTICLE EMISSION

A nucleus which undergoes radioactive decay by emitting an α-particle is able to do so because its mass is greater than the sum of the masses of the daughter nucleus and the emitted α-particle.*

*Fulfilment of this condition does not guarantee that an α-particle is emitted.

In order to illustrate this we shall consider $^{210}_{84}\text{Po}$, which decays according to

$$^{210}_{84}\text{Po} \longrightarrow {}^{206}_{82}\text{Pb} + {}^{4}_{2}\text{He}$$

The various terms in this equation can be taken to represent <u>atoms</u> rather than nuclei because each side of the equation involves the same number of electrons. From tables

$$\text{Atomic mass of } {}^{210}_{84}\text{Po} = 209.983\,\text{u}$$

$$\text{Atomic mass of } {}^{206}_{82}\text{Pb} = 205.974\,\text{u}$$

$$\text{Atomic mass of } {}^{4}_{2}\text{He} = 4.003\,\text{u}$$

The mass of the parent (209.983 u) is greater than the total mass of the products $(205.974 + 4.003 = 209.977\,\text{u})$ – as required.

It can be shown that about 98% of the energy provided by the decrease in mass is carried away as the kinetic energy of the α-particle; the remaining 2% is the recoil energy of the nucleus. (The nucleus recoils in order that momentum is conserved.)

2.5 STABILITY AGAINST (NEGATIVE) β-PARTICLE EMISSION

If a nucleus is to decay by emitting a β-particle, the mass of the nucleus must be greater than the total mass of the decay products. Consider the possibility of $^{14}_{6}\text{C}$ decaying by (negative) β-emission. If it does, the relevant <u>nuclear</u> equation is

$$^{14}_{6}\text{C} \longrightarrow {}^{14}_{7}\text{N} + {}^{0}_{-1}\text{e}$$

Adding six electrons to each side of the equation gives

$$^{14}_{6}\text{C} + 6{}^{0}_{-1}\text{e} \longrightarrow {}^{14}_{7}\text{N} + 7{}^{0}_{-1}\text{e}$$

Bearing in mind that a carbon atom has six electrons and that a nitrogen atom has seven electrons, we can rewrite this equation as

$$^{14}_{6}\text{C} \longrightarrow {}^{14}_{7}\text{N}$$

where the terms now represent <u>atoms</u> rather than nuclei. From tables:

$$\text{Atomic mass of } {}^{14}_{6}\text{C} = 14.003\,2\,\text{u}$$

$$\text{Atomic mass of } {}^{14}_{7}\text{N} = 14.003\,1\,\text{u}$$

Thus the atomic mass of $^{14}_{7}\text{N}$ is less than that of $^{14}_{6}\text{C}$ and the decay is possible, and in fact <u>does</u> occur.

EXAMPLE 2.2

Calculate the energy released (the **Q-value**) when gallium 70 ($^{70}_{31}\text{Ga}$) undergoes (negative) β-decay to produce germanium 70 ($^{70}_{32}\text{Ge}$). (Atomic mass of $^{70}_{31}\text{Ga} = 69.926\,05\,\text{u}$, of $^{70}_{32}\text{Ge} = 69.924\,25\,\text{u}$. $1\,\text{u} = 931.5\,\text{MeV}$.)

Solution

The <u>nuclear</u> equation is

$$^{70}_{31}\text{Ga} \longrightarrow {}^{70}_{32}\text{Ge} + {}^{0}_{-1}\text{e}$$

Adding 31 electrons to each side of the equation gives

$$^{70}_{31}\text{Ga} \longrightarrow {}^{70}_{32}\text{Ge}$$

where the terms now represent atoms.

Decrease in mass* $= 69.926\,05 - 69.924\,25 = 0.001\,80\,\text{u}$

$\therefore$ Energy released $= 0.001\,80 \times 931.5 = 1.68\,\text{MeV}$

2.6 STABILITY AGAINST POSITRON EMISSION (POSITIVE β-PARTICLE EMISSION)

A number of artificially produced radioactive nuclides decay by emitting a **positron**. (A positron is a positively charged particle which has the same mass as the electron, and whose charge is numerically equal to that of the electron.) For example, α-particle bombardment of $^{19}_{9}\text{F}$ produces $^{22}_{11}\text{Na}$ – an artificial isotope of sodium. This decays by positron emission (sometimes called positive β-decay) to produce $^{22}_{10}\text{Ne}$ – a stable isotope of neon. The reactions are:

$$^{19}_{9}\text{F} + {}^{4}_{2}\alpha \rightarrow {}^{22}_{11}\text{Na} + {}^{1}_{0}\text{n}$$

and

$$^{22}_{11}\text{Na} \rightarrow {}^{22}_{10}\text{Ne} + {}^{0}_{1}\text{e}$$
$$\text{(Positron)}$$

There are no naturally occurring positron emitters. Decay is possible only if the combined mass of the positron and the daughter nucleus is less than that of the parent. See Questions 2C number 2.

2.7 STABILITY AGAINST FISSION

The fission process is discussed in section 2.10. The ideas of the last three sections apply, i.e. the total mass of the fission products is less than that of the nucleus which has undergone fission.

QUESTIONS 2C

1. Radium 224 decays by α-emission to produce radon 220 according to

    $$^{224}_{88}\text{Ra} \longrightarrow {}^{220}_{86}\text{Rn} + {}^{4}_{2}\text{He}$$

    Calculate: **(a)** the decrease in mass, **(b)** the energy released (the **Q-value**).
    (Atomic mass of $^{224}_{88}\text{Ra} = 224.020\,22\,\text{u}$, of $^{220}_{86}\text{Rn} = 220.011\,40\,\text{u}$, of $^{4}_{2}\text{He} = 4.002\,60\,\text{u}$. $1\,\text{u} = 931.5\,\text{MeV}$.)

2. Nitrogen 13 decays by positron emission to produce carbon 13. The relevant <u>nuclear</u> equation is

    $$^{13}_{7}\text{N} \longrightarrow {}^{13}_{6}\text{C} + {}^{0}_{1}\text{e}$$

    Calculate: **(a)** the decrease in mass, **(b)** the energy released (the **Q-value**).
    (Atomic mass of $^{13}_{7}\text{N} = 13.005\,74\,\text{u}$, of $^{13}_{6}\text{C} = 13.003\,35\,\text{u}$. Mass of $^{0}_{1}\text{e}$ and of $^{0}_{-1}\text{e} = 0.000\,55\,\text{u}$. $1\,\text{u} = 931.5\,\text{MeV}$.) Hint – add 7 electrons to each side of the nuclear equation.

3. A nucleus decays to produce an α-particle of mass $4.00\,\text{u}$ and a daughter of mass $204\,\text{u}$, releasing $5.21\,\text{MeV}$ in the process. **(a)** Bearing in mind that the daughter nucleus recoils in order that momentum is conserved, calculate the value of the ratio: kinetic energy of α-particle/kinetic energy of daughter. **(b)** Hence find the kinetic energy of the α-particle.

*This should not be referred to as the 'mass defect'.

2.8 STABILITY AND NEUTRON–PROTON RATIO

Fig. 2.2 shows a plot of neutron number against proton number (a **Segrè chart**) for all known stable nuclides. It can be seen that among the light nuclei the tendency is for there to be equal numbers of neutrons and protons. Heavy nuclei, on the other hand, have more neutrons than protons, the neutron–proton ratio reaching about 1.5 with $^{208}_{82}$Pb.

Fig. 2.2
Variation of neutron number with proton number for stable nuclides

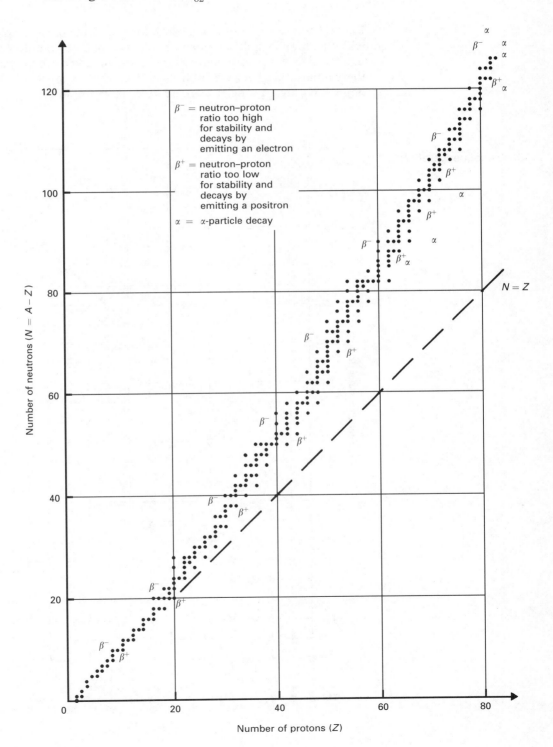

Nuclides which lie above the stability belt (in the region marked β^-) have a neutron–proton ratio which is too high for stability and are likely to decay by emitting a (negative) β-particle, i.e. by converting a neutron to a proton. Nuclides which lie below the stable region (marked β^+) have a neutron–proton ratio which is too low and can correct this by converting a proton to a neutron (i.e. by emitting a positron) or, when $Z > 60$, by emitting an α-particle. All nuclides with $Z > 83$ are unstable; those that occur in nature decay either by α-emission or β^--emission.

Fig. 2.3 shows a plot of N against Z for the whole of the uranium series together with that for all stable nuclides with $N \geq 118$. The plot illustrates very effectively that the decay sequence stays more or less on course for, and eventually reaches, the stability belt because both α- and β^--decays are involved. This would not happen if only α-particles were emitted for example.

Fig. 2.3
Decay sequence of $^{238}_{92}$U

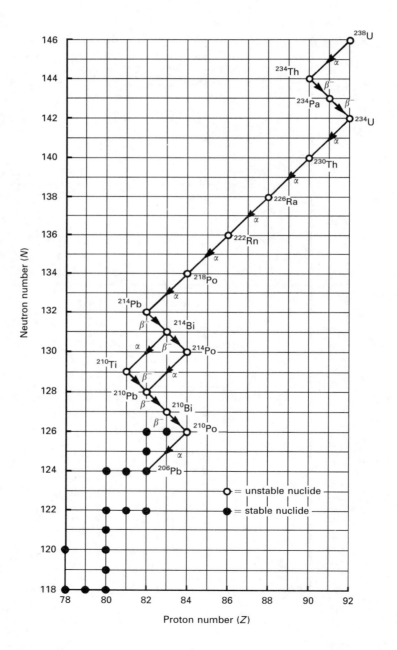

2.9 NUCLEON EMISSION

Some particularly unstable nuclides can decay by emitting a nucleon.

Neutron Emission

Neutron emitters lie above the stability belt (Fig. 2.2) and in most cases can also decay by β^--emission. There are no naturally occurring neutron emitters, most of them are produced by β^--decay of fission products in nuclear reactors (see section 2.10). They lie along the upper edge of the region marked β^- in Fig. 2.2 and have large excesses of neutrons. An example is yttrium 99, which has ten more neutrons than the <u>stable</u> isotope of yttrium – yttrium 89.

Proton Emission

Proton emitters lie along the bottom edge of the region marked β^+ in Fig. 2.2. An example is neon 17 with three fewer neutrons than the most abundant isotope of neon – neon 20. There are no natural proton emitters.

2.10 NUCLEAR FISSION

Nuclear fission is the disintegration of a heavy nucleus into two lighter nuclei. Energy is released by the process because the average binding energy per nucleon of these lighter nuclei is greater than that of the parent.

As an example of a fission reaction, consider the bombardment of $^{235}_{92}U$ by <u>slow</u> neutrons. This can result in the capture of a neutron and the formation of $^{236}_{92}U$ which is in a highly excited state and undergoes fission almost immediately. Many different pairs of nuclei (known as **primary fission fragments**) can be produced by the fission of $^{235}_{92}U$, but in over 95% of the cases the mass number of the heavier fragment is between 130 and 149. These fragments have too many neutrons to be stable and emit an average of 2.5 neutrons, called **prompt neutrons**, immediately after fission has occurred. One possible reaction, in which three prompt neutrons are released by the primary fission fragments to produce $^{146}_{57}La$ and $^{87}_{35}Br$, is

$$^{235}_{92}U + {}^{1}_{0}n \rightarrow {}^{236}_{92}U \rightarrow {}^{146}_{57}La + {}^{87}_{35}Br + 3\,{}^{1}_{0}n + \text{Energy}$$

What we shall refer to as the **primary fission products** ($^{146}_{57}La$ and $^{87}_{35}Br$ in this case) are also neutron-rich and usually decay by way of several negative β-emissions to a stable end product. The primary fission products and the various nuclides in the decay chains they initiate are collectively referred to as **fission products**.

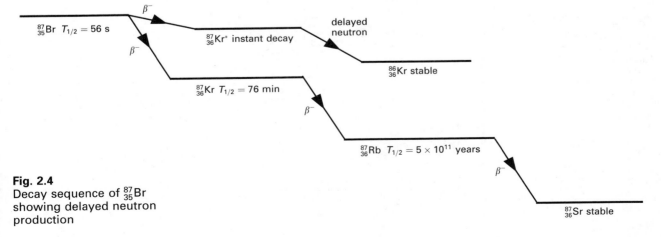

Fig. 2.4
Decay sequence of $^{87}_{35}Br$ showing delayed neutron production

Occasionally β^--decay of a fission product produces a daughter nucleus which is in such an excited state that it decays by emitting a neutron, known as a **delayed neutron** because it may be produced several seconds after the prompt neutrons (see Fig. 2.4). Delayed neutrons account for only about 0.7% of the total number of neutrons produced.

Nuclear reactors (section 5.2) make use of controlled fission reactions to provide energy. The **atom bomb** makes use of an uncontrolled fission reaction.

Fig. 2.5
Distribution of primary fission products of $^{235}_{92}$U

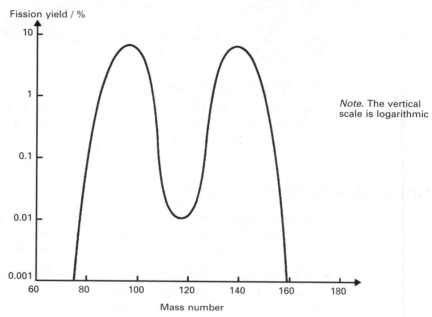

Note. The vertical scale is logarithmic

Notes (i)   Fig. 2.5 shows the relative proportions of the various primary fission products produced by the fission of $^{235}_{92}$U.

(ii)   Fig. 2.6 illustrates how the energy released in fission is accounted for in terms of the plot of binding energy per nucleon against mass number.

(iii) The primary fission fragments (and many of the various fission products) are unstable because their neutron–proton ratios are too high. This can be appreciated by examining Fig. 2.2. We see that neutron–proton ratio increases with mass number. Since there is no change in the total number of neutrons and protons, and therefore no change in the overall neutron–proton ratio, at least one of the primary fission fragments is bound to have a neutron–proton ratio that puts it above the stability belt.

Fig. 2.6
To illustrate release of energy in fission of a heavy nucleus

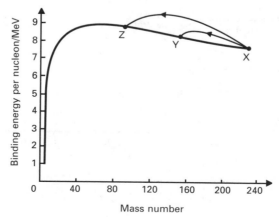

$$X \rightarrow Y + Z$$

The primary fission fragments (Y and Z) have greater binding energy per nucleon than the nucleus undergoing fission (X). Since there is no change in the total number of nucleons, there is an increase in binding energy, i.e. a release of energy.

EXAMPLE 2.3

Calculate the energy released when 10 kg of $^{235}_{92}$U undergoes fission according to

$$^{235}_{92}\text{U} + ^1_0\text{n} \longrightarrow ^{141}_{56}\text{Ba} + ^{92}_{36}\text{Kr} + 3^1_0\text{n}$$

(Mass of $^{235}_{92}$U = 235.04 u, of $^{141}_{56}$Ba = 140.91 u, of $^{92}_{36}$Kr = 91.91 u, of $^1_0$n = 1.01 u, 1 u = 931.5 MeV, N_A = 6.02×10^{23} mol^{-1}.)

Solution

Mass difference = $(235.04 + 1.01) - (140.91 + 91.91 + 3 \times 1.01)$

= 0.20 u

$\therefore$ Energy released = 0.20×931.5 = 186.3 MeV

235.04×10^{-3} kg of $^{235}_{92}$U contains 6.02×10^{23} atoms

$\therefore$ 10 kg of $^{235}_{92}$U contains $\dfrac{10 \times 6.02 \times 10^{23}}{235.04 \times 10^{-3}}$ = 2.56×10^{25} atoms

$\therefore$ Energy released by 10 kg of $^{235}_{92}$U = $2.56 \times 10^{25} \times 186.3$

= 4.8×10^{27} MeV

EXAMPLE 2.4

Calculate the energy released (i.e. the **Q-value**) when a uranium 236 nucleus undergoes fission according to

$$^{236}_{92}\text{U} \longrightarrow ^{146}_{57}\text{La} + ^{87}_{35}\text{Br} + 3^1_0\text{n}$$

(Binding energy per nucleon of $^{236}_{92}$U = 7.59 MeV, of $^{146}_{57}$La = 8.41 MeV, of $^{87}_{35}$Br = 8.59 MeV.)

Solution

Binding energy of $^{146}_{57}$La = 146×8.41 = 1227.86 MeV

Binding energy of $^{87}_{35}$Br = 87×8.59 = 747.33 MeV

$\therefore$ Total binding energy after fission = $1227.86 + 747.33$ = 1975.19 MeV

Binding energy of $^{236}_{92}$U = 236×7.59 = 1791.24 MeV

$\therefore$ Increase in binding energy = $1975.19 - 1791.24$ = 183.95 MeV

$\therefore$ Energy released = 184 MeV

2.11 NUCLEAR FUSION

Nuclear fusion is the combining of two light nuclei to produce a heavier nucleus. Energy is released by the process. An example is the fusion of two deuterium nuclei to produce helium 3:

$$^2_1\text{H} + ^2_1\text{H} \rightarrow ^3_2\text{He} + ^1_0\text{n} + 3.27\,\text{MeV}^\star$$

$^\star{}^2_1\text{H} + ^2_1\text{H} \rightarrow ^3_1\text{H} + ^1_1\text{p} + 4.03$ MeV occurs with equal probability.

The nuclei which are to fuse need to approach each other closely enough for the strong interaction to bind them together. In order to do this they must first overcome their mutual electrostatic repulsion. One way of giving them the necessary energy is to heat them to a temperature of the order of 10^8 K. At temperatures as high as this all the atoms involved (all of which are <u>light</u>) will have lost all their electrons, creating a completely ionized gas known as a **plasma** – the so-called fourth state of matter. Nuclear fusion brought about in this manner is called **thermonuclear fusion**.

To date, no one has succeeded in producing a <u>controlled</u> thermonuclear fusion reaction which has been self-sustaining. The **thermonuclear bomb** (the **hydrogen bomb**) makes use of an <u>uncontrolled</u> reaction – the high temperature required for the fusion to occur is provided by the explosion of an atom (i.e. a fission) bomb.

The energy released by the fusion of two nuclei is very much less than that which results from the fission of, say, a uranium nucleus. However, it should be borne in mind that fusion involves very much less massive nuclei and in fact, some fusion reactions provide over four times as much energy <u>per unit mass of reactants</u> as that produced by fission.

Notes   (i)   The kinetic theory gives the average kinetic energy of the particles of a gas at a kelvin temperature T as $\frac{3}{2}kT$, where k is Boltzmann's constant. If we apply this result to plasma particles at 10^8 K, we find that their average kinetic energy is 2.07×10^{-15} J $= 12.9$ keV. The electrostatic potential energy of, for example, two deuterium nuclei at a separation of 4×10^{-15} m (where the strong interaction might be expected to start to take over from electrostatic repulsion) is about 360 keV. This is considerably more than 12.9 keV and at first sight it appears that temperatures much in excess of 10^8 K are required if fusion is to occur. There are two reasons why a temperature of around 10^8 K is sufficient.

(a)   12.9 keV is the <u>average</u> kinetic energy at 10^8 K, the particles on the high-energy tail of the Maxwell velocity distribution would be much more energetic than this.

(b)   There is the possibility of some particles tunnelling (see section 4.5) through the potential barrier.

(ii)   It is possible to produce fusion by causing particles to collide in, for example, a linear accelerator. However, this consumes much more energy than it produces and therefore cannot be used as a means of generating power.

QUESTIONS 2D

1.   Calculate the energy released (the **Q-value**) by the fusion reaction

$$^6_3\text{Li} + {}^2_1\text{H} \rightarrow {}^7_3\text{Li} + {}^1_1\text{H}$$

(Atomic mass of $^6_3\text{Li} = 6.015\,13$ u, of $^2_1\text{H} = 2.014\,10$ u, of $^7_3\text{Li} = 7.016\,01$ u and of $^1_1\text{H} = 1.007\,83$ u, 1 u $= 931.5$ MeV.)

2.   How much energy would be released if 1.00 kg of deuterium $\left({}^2_1\text{H}\right)$ were to undergo fusion by the reaction in Question 1? (The Avogadro constant, $N_\text{A} = 6.02 \times 10^{23}$ mol^{-1}.)

CONSOLIDATION

The electronvolt (eV) is a unit of energy equal to the kinetic energy gained by an electron in being accelerated through a PD of 1 volt.

$$E = mc^2 \qquad 1\,u = 931.5\,\text{MeV}$$

$$\text{Mass defect of an atom} = \left(\begin{array}{c}\text{Mass of neutrons} \\ \text{protons and electrons}\end{array}\right) - \left(\begin{array}{c}\text{Mass of} \\ \text{atom}\end{array}\right)$$

The binding energy of an atom is the energy required to split it up into its component neutrons, protons and electrons.

$$\underset{\text{(MeV)}}{\text{Binding energy}} = 931.5 \times \underset{\text{(u)}}{\text{Mass defect}}$$

$$\underset{\text{(J)}}{\text{Binding energy}} = \underset{\text{(kg)}}{\text{Mass defect}} \times \underset{(\text{m s}^{-1})^2}{c^2}$$

Fission The disintegration of a heavy nucleus into two lighter ones, accompanied by a release of energy and an associated decrease in mass.

The plot of binding energy per nucleon against mass number (Fig. 2.1) shows that there is an increase in binding energy, and therefore a release of energy, when two light nuclei fuse to produce a heavier one, and when a heavy nucleus undergoes fission producing two lighter ones.

$$\text{Energy released} = \text{Increase in binding energy}$$

$$\underset{\text{(MeV)}}{\text{Energy released}} = 931.5 \times \underset{\text{(u)}}{\text{Decrease in mass}}$$

$$\underset{\text{(J)}}{\text{Energy released}} = \underset{\text{(kg)}}{\text{Decrease in mass}} \times \underset{(\text{m s}^{-1})^2}{c^2}$$

To calculate the energy released in α-decay When working with <u>atomic</u> masses the terms in the nuclear equation can be taken to represent atoms.

To calculate the energy released in β-decay When working with <u>atomic</u> masses remember to add a suitable number of electrons to each side of the nuclear equation.

Fusion The joining of two light nuclei to produce a heavier one, accompanied by a release of energy and an associated decrease in mass.

Thermonuclear fusion A fusion reaction in which the nuclei that fuse have sufficient kinetic energy to overcome their mutual electrostatic repulsion because they are at a high temperature ($\sim 10^8$ K).

The thermonuclear bomb (the hydrogen bomb) makes use of an <u>uncontrolled</u> thermonuclear reaction.

QUESTIONS ON CHAPTER 2

1. (a) Explain the meaning of the term *mass difference* and state the relationship between the mass difference and the *binding energy* of a nucleus.

(b) Sketch a graph of nuclear binding energy per nucleon versus mass number for the naturally occurring isotopes and show how it may be used to account for the possibility of energy release by nuclear fission and by nuclear fusion.

(c) The Sun obtains its radiant energy from a thermonuclear fusion process. The mass of the Sun is 2×10^{30} kg and it radiates 4×10^{23} kW at a constant rate. Estimate the lifetime of the Sun, in years, if 0.7% of its mass is converted into radiation during the fusion process and it loses energy only by radiation. (1 year may be taken as 3×10^7 s.)

(The speed of light, $c = 3 \times 10^8 \, \text{m s}^{-1}$.)   [J]

2. (a) Sketch a graph to show how the number of neutrons, N, varies with the number of protons, Z, for naturally occurring stable and unstable nuclei over the range $Z = 0$ to $Z = 90$. On the graph, show values on the N axis, draw the line $N = Z$, and indicate clearly the region in which stable nuclides would occur.

On the same graph, mark points, one for each, to indicate the position of an unstable nuclide which would be likely to be
   (i) an α-emitter, labelling it A,
   (ii) a β^--emitter, labelling it B,
   (iii) a β^+-emitter, labelling it C.

(b) State the changes in N and Z which are produced in (i) α, (ii) β^-, and (iii) β^+ emission. Hence explain why the nuclides A, B and C lie in the regions you have indicated in (a) above.

(c) The nuclide $^{210}_{84}$Po is an alpha-emitter. Show that the emitted alpha particle has 98% of the released kinetic energy. Without further calculation, compare this with the sharing of kinetic energy which takes place in beta-decay.   [J, '92]

3. (a) Explain what is meant by the *mass defect* of an atom.

(b) An isotope of lead $^{210}_{82}$Pb has an atomic mass of 209.984 u. Calculate the mass defect (in u) of an atom of this isotope.

(The mass of a neutron is 1.009 u, and that of a hydrogen atom (= 1 proton + 1 electron) is 1.008 u.)

(c) (i) What is the significance of the binding energy per nucleon of an isotope?

(ii) How can this quantity be found from the mass defect?   [O, '92]

4.

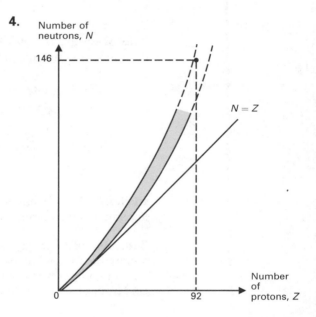

N, Z values for the stable nuclides lie within the shaded area of the figure.

(a) A part of the naturally-occurring uranium radioactive series is given by

$$^{238}_{92}\text{U} \xrightarrow[\alpha, \, \beta^-, \, \beta^-, \, \alpha, \, \alpha, \, \alpha]{} \text{X}$$

where α represents decay by emission of an alpha-particle and β^- represents decay by emission of a negative beta-particle. The decay processes occur in the order shown under the arrow.

(i) Calculate the proton number and the neutron number of the nuclide X.

(ii) With reference to the figure above, explain why there are both α and β^- emitters in the series.

(b) The uranium series discussed in part (a) continues from the nuclide X through several more decay stages, concluding with the nuclide $^{206}_{82}$Pb.

Calculate the binding energy per nucleon of **(i)** the $^{238}_{92}$U nucleus, **(ii)** the $^{206}_{82}$Pb nucleus. Comment on your results.

(c) What happens to the energy released at each stage of the decay series?

Mass of an electron $= 0.000\,55\,u$
Mass of a proton $= 1.007\,28\,u$
Mass of a neutron $= 1.008\,67\,u$
Mass of a neutral atom of $^{206}_{82}Pb = 205.974\,46\,u$
Mass of a neutral atom of $^{238}_{92}U = 238.050\,76\,u$

Masses are quoted in unified atomic mass units, u.
1 u is equivalent to 931 MeV.   [J]

5. Data to be used in this question should be selected from the table opposite.
The figure shows the neutron number, N, plotted against the proton number, Z, for the *stable* nuclides up to $Z = 20$.

(a) Use the figure to decide which of the nuclei $^{13}_{7}N$ and $^{34}_{15}P$ is a β^--emitter and which is a β^+-emitter. Give reasons for your answer, and write equations for both processes.

(b) Calculate the kinetic energy released, in MeV, in each case. Explain how the masses of the electrons are accounted for in the equations.

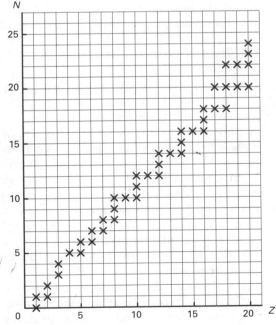

In the following table, which lists a selection of nuclides, the mass of an isotope is given for a neutral atom of the substance and is quoted in unified atomic mass units, u.

(1 u is equivalent to 931.3 MeV, mass of an electron $= 0.000\,55\,u$, mass of a positron $= 0.000\,55\,u$.)

Name	Symbol	Proton number, Z	Nucleon number, A	Atomic mass/u
Proton	p	1	1	1.00728
Neutron	n	0	1	1.00867
Hydrogen	^1H	1	1	1.00783
	^2H	1	2	2.01410
	^3H	1	3	3.01605
Helium	^3He	2	3	3.01603
	^4He	2	4	4.00260
Carbon	^{11}C	6	11	11.01143
	^{12}C	6	12	12.00000
	^{13}C	6	13	13.00335
Nitrogen	^{13}N	7	13	13.00574
	^{14}N	7	14	14.00307
	^{15}N	7	15	15.00011
Oxygen	^{13}O	8	13	13.01200
	^{14}O	8	14	14.00860
	^{15}O	8	15	15.00307
	^{16}O	8	16	15.99492
Silicon	^{28}Si	14	28	27.97693
	^{29}Si	14	29	28.97649
	^{30}Si	14	30	29.97376
	^{33}Si	14	33	32.97400
	^{34}Si	14	34	33.97450
Phosphorus	^{30}P	15	30	29.97832
	^{31}P	15	31	30.97376
	^{32}P	15	32	31.97391
	^{33}P	15	33	32.97173
	^{34}P	15	34	33.97330
Sulphur	^{31}S	16	31	30.97960
	^{32}S	16	32	31.97207
	^{33}S	16	33	32.97146
	^{34}S	16	34	33.96786
	^{35}S	16	35	34.96903

[J]

6. Sketch a graph showing the variation of the number of neutrons N with the number of protons Z for stable nuclei. Indicate approximate numerical values of N and Z on the axes and draw the line $N = Z$.
By referring to the graph, explain
(a) why, in a *natural* radioactive series, radioactive decays involving α-emissions and β^--emissions both occur,
(b) why, in the series in (i), β^+-emissions are unlikely,
(c) what would be a likely form of radioactivity of the fission fragments resulting from a nuclear fission.   [J]

7. Explain the term *nuclear binding energy*. Sketch a graph showing the variation of binding energy per nucleon number (mass number) and show how both nuclear fission and nuclear fusion can be explained from the shape of this curve.

Calculate in MeV the energy liberated when a helium nucleus ($^4_2$He) is produced **(a)** by fusing two neutrons and two protons, and **(b)** by fusing two deuterium nuclei ($^2_1$H). Why is the quantity of energy different in the two cases? (The neutron mass is 1.008 98 u, the proton mass is 1.007 59 u, the nuclear masses of deuterium and helium are 2.014 19 u and 4.002 77 u respectively. 1 u is equivalent to 931 MeV.)   [L]

8. When $^{235}_{92}$U is bombarded by neutrons, two possible fission products are $^{87}_{35}$Br and $^{146}_{57}$La. Give an equation for this process and calculate the energy released in MeV.

Atomic mass of $^{235}_{92}$U $= 235.04$ u
Atomic mass of $^{87}_{35}$Br $= 86.92$ u
Atomic mass of $^{146}_{57}$La $= 145.90$ u
Mass of $^1_0$n $= 1.01$ u
1 u is equivalent to 931 MeV.   [J]

9. In the fusion reaction $^2_1$H $+ ^3_1$H $= ^4_2$He $+ ^1_0$n, how much energy, in joules, is released? (Mass of $^2_1$H $= 3.345 \times 10^{-27}$ kg, of $^3_1$H $= 5.008 \times 10^{-27}$ kg, of $^4_2$He $= 6.647 \times 10^{-27}$ kg, of $^1_0$n $= 1.675 \times 10^{-27}$ kg, speed of light $= 3.0 \times 10^8$ m s^{-1}.)   [L]

10. (a) State the changes to the number of protons and of neutrons that occur within nuclei when they emit
    (i) α-particles,
    (ii) β-particles,
    (iii) γ-radiation.
(b) A sample of radioactive material emits a narrow parallel beam of α-particles, β-particles and γ-radiation as illustrated in the figure below.

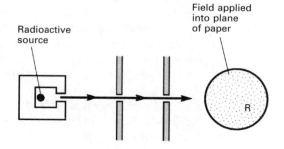

The beam passes through a region R where a uniform field may be applied at right angles to the beam and into the plane of the paper. Discuss the effect on the beam if the field in the region R is

    (i) an electric field,
    (ii) a magnetic field.
(c) When beryllium is bombarded with α-particles of energy 8.0×10^{-13} J, carbon atoms are produced, together with a very penetrating radiation. The nuclear reaction might be

Either   **(i)** $^9_4$Be $+ ^4_2$He $\longrightarrow ^{13}_6$C $+ \gamma$

Or   **(ii)** $^9_4$Be $+ ^4_2$He $\longrightarrow ^{12}_6$C $+ ^1_0$n

1 Explain what is meant by $^{13}_6$C

2 The energy of the penetrating radiation is found to be at least 8.8×10^{-12} J for each γ or $^1_0$n produced. Show that equation **(c)(i)** cannot be valid. Explain your reasoning carefully.

Nuclide	Mass/u
$^9_4$Be	9.0150
$^4_2$He	4.0040
$^{13}_6$C	13.0075

[C, '93]

11. Describe briefly how **(a)** the nuclear binding energy per nucleon, and **(b)** the neutron/proton ratio vary with the nucleon number A of stable nuclei.

Hence explain the following observations:
(c) Elements of very high A do not occur naturally.
(d) High A elements frequently decay by α-emission.
(e) β^--particles often accompany such an emission of α-particles, whereas β^+-particles seldom do so.

A particular isotope X ($A = 220$) decays by emission of an α-particle together with a γ-ray to isotope Y. The half-life of this decay is 2000 years. Isotope Y then decays by β^--emission to a stable isotope Z with a half-life of 10 days. A sample of mass 0.1 mg of pure isotope X is prepared on June 1st, and the β-activity of the sample is measured throughout the following summer.
(f) Describe how the experiment might be carried out.
(g) Calculate the number of β-particles emitted per second by the sample on September 1st, and the mass of isotope Y present at this date.
(h) Discuss how you expect the β-activity to vary with time over the month of June. How would the activity on June 11th relate to that on September 1st?
($N_A = 6.0 \times 10^{23}$ mol^{-1}.)

[O & C (special)]

12. The first example of an induced nuclear transformation detected in the laboratory resulted from the bombardment of nitrogen by alpha-particles:

$$^{14}_{7}N + \alpha \rightarrow ^{17}_{8}O + p$$

Use the following data to calculate the increase in rest mass which occurs: mass of nitrogen 14 nucleus $= 23.2466 \times 10^{-27}$ kg, mass of oxygen 17 nucleus $= 28.2209 \times 10^{-27}$ kg, mass of alpha-particle $= 6.6442 \times 10^{-27}$ kg, mass of proton $= 1.6725 \times 10^{-27}$ kg. State the source of this increased rest mass. Calculate the minimum kinetic energy of the incident alpha-particle for the transformation to be possible.

(Speed of light in vacuum, $c = 3.0 \times 10^8$ m s^{-1}.)

[L, '94]

3

GROSS PROPERTIES OF NUCLEI

3.1 THE 'PLUM-PUDDING' MODEL OF THE ATOM

By 1900, the idea that matter was composed of atoms was well established. It was known that atoms contained negatively charged particles (**electrons**). It was also known that atoms as a whole are electrically neutral, and therefore they must have a positively charged component. No positively charged equivalent of an electron was known (positive ions are much more massive), and this led J. J. Thomson to propose the so called **'plum-pudding' model** of the atom. In this model, the positive charge was supposed to be continuously distributed throughout a sphere in which the electrons were embedded – rather like plums in a pudding. The model had very limited success and, in particular, was totally incapable of accounting for the results of Rutherford's α-particle scattering experiments.

3.2 THE SCATTERING OF α-PARTICLES

In 1909 Rutherford investigated the scattering of α-particles (positively charged particles resulting from radioactive decay) by thin films of heavy metals, notably gold. (The experiments were actually carried out by Geiger and Marsden under Rutherford's direction.) The experimental arrangement is shown in Fig. 3.1.

Fig. 3.1
Rutherford's apparatus to investigate scattering of α-particles

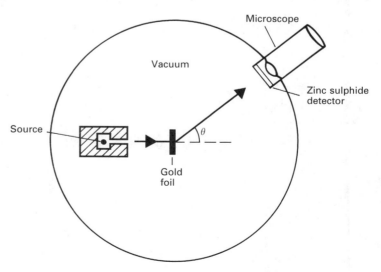

A narrow pencil of α-particles from a radon source inside a metal block was incident on a thin metal foil. A glass screen coated with zinc sulphide was used to detect the scattered α-particles. Whenever a particle hit the screen it produced a faint flash of light (a **scintillation**). The experiment was carried out in a darkened room and the scintillations were observed through a microscope. The screen could be rotated about the metal foil, and by counting the number of scintillations produced in various positions in equal intervals of time, the angular dependence of the scattering was determined. The range of α-particles in air is limited to about 5 cm, and therefore the apparatus was evacuated so that the particles would not be prevented from reaching the screen.

Ernest Rutherford
(1871–1937)

The majority of the α-particles were scattered through small angles, but a few (about 1 in 8000) were deviated by more than 90°.

The most popular model of the atom at the time was the plum-pudding model. The net charge at any point in such an atom is small, and therefore it would not be able to scatter an α-particle through a large angle. The large deflections which were observed might be thought to be due to successive deflections through small angles. However, Rutherford was able to show that the number of large deflections, though small, was far too high to be accounted for in this way, and he suggested that the large-angle scattering was due to a single encounter between an α-particle and an intense positive electric charge. In view of this, Rutherford proposed in 1911 that **an atom has a positively charged core (now called the nucleus) which contains most of the mass of the atom and which is surrounded by orbiting electrons**. On the basis of this model, Rutherford calculated that the number of particles scattered through an angle, θ, should be proportional to $\operatorname{cosec}^4 (\theta/2)$. Geiger and Marsden performed a second series of experiments and verified this prediction in 1913.

Large-angle scattering occurred whenever an α-particle was incident almost head-on to a nucleus (see Fig. 3.2). Since very few of the particles were scattered through large angles, it follows that the probability of a head-on approach is small and indicated that **the nucleus occupies only a small proportion of the available space**. (The nuclear radius is of the order of 10^{-15} m; that of an atom as a whole is about 10^{-10} m.)

Fig. 3.2
Scattering of α-particles
by a nucleus

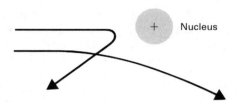

3.3 ESTIMATE OF NUCLEAR RADIUS BY α-PARTICLE SCATTERING

Consider an α-particle incident <u>head-on</u> to a nucleus (Fig. 3.3). As the α-particle approaches the nucleus its kinetic energy decreases and its electrostatic potential energy (due to Coulomb repulsion from the positively charged nucleus) increases.

Fig.3.3
To calculate closest
distance of approach of
α-particle

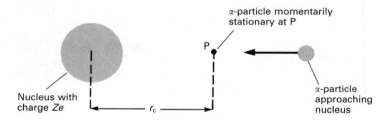

Suppose that P is the point at which the α-particle is <u>closest</u> to the nucleus, a distance r_c from its centre. The kinetic energy of the α-particle is momentarily zero at P, and therefore if its initial kinetic energy is E_α, then in moving to P

$$\text{KE lost by } \alpha\text{-particle} = E_\alpha$$

Initially the α-particle is far from the nucleus where its potential energy can be taken to be zero, and therefore in moving to P

$$\text{PE gained by } \alpha\text{-particle} = \frac{1}{4\pi\varepsilon_0} \cdot \frac{Ze \times 2e}{r_c} \quad \left[= \left(\begin{array}{c} \text{potential at P} \\ \text{due to nucleus} \end{array} \right) \times \left(\begin{array}{c} \text{charge on} \\ \alpha\text{-particle} \end{array} \right) \right]$$

If we make the reasonable assumption that the nucleus remains at rest throughout,

$$\text{KE lost by } \alpha\text{-particle} = \text{PE gained by } \alpha\text{-particle}$$

$$\therefore \quad E_\alpha = \frac{1}{4\pi\varepsilon_0} \cdot \frac{Ze \times 2e}{r_c}$$

i.e. $$r_c = \frac{2Ze^2}{4\pi\varepsilon_0 E_\alpha} \qquad\qquad [3.1]$$

In Rutherford's experiment:

$$E_\alpha = 7.68\,\text{MeV} = 7.68 \times 10^6 \times 1.6 \times 10^{-19} = 1.23 \times 10^{-12}\,\text{J}$$

$$Z = 79 \text{ (the atomic number of gold)}$$

Also

$$\varepsilon_0 = 8.85 \times 10^{-12} \, \text{F m}^{-1}$$

$$e = 1.60 \times 10^{-19} \, \text{C}$$

Substituting these values into equation [3.1] gives

$$r_c = 2.96 \times 10^{-14} \, \text{m}$$

In deriving equation [3.1], like Rutherford, we have assumed that the potential at P is due to a point charge at the centre of the nucleus. This is justified providing P is outside the nucleus; it would not be if P were inside the nucleus. Rutherford also obtained an expression for the number $N(\theta)$ of α-particles scattered through any given angle θ – again on the basis of the nucleus acting as a point charge. His experimental values of $N(\theta)$ were in excellent agreement with the predicted values in the case of 7.68 MeV α-particles scattered by gold nuclei. This meant that the α-particles had not actually penetrated the nucleus and therefore the radius of a gold nucleus must be less than $2.96 \times 10^{-14} \, \text{m}$.

α-particles of higher energy approach the nucleus more closely, and so give smaller values of r_c. If α-particles of increasingly higher energies are used, a point is reached at which the experimentally observed values of $N(\theta)$ deviate from the theoretical values, indicating that these α-particles have just penetrated the nucleus. The corresponding value of r_c can be taken to be the radius of the nucleus. Experiments of this kind, using a wide variety of target nuclei, reveal that the radius R of any nucleus can be represented by

$$R = r_0 A^{1/3} \tag{3.2}$$

where

$$r_0 = \text{a constant} \, (= 1.414 \times 10^{-15} \, \text{m})$$

$$A = \text{the mass number of the nucleus}$$

Notes   (i)   Nuclear radii are commonly expressed in **femtometres** (fm).

$$1 \, \text{fm} = 1 \times 10^{-15} \, \text{m}$$

The femtometre is often called a **fermi** (in honour of Enrico Fermi). The fermi is not an SI unit.

(ii)   Experiments using projectiles other than α-particles (e.g. electrons, neutrons and protons) give slightly different values of r_0.

3.4 ESTIMATE OF NUCLEAR RADIUS BY ELECTRON DIFFRACTION (HIGH-ENERGY ELECTRON SCATTERING)

Electron diffraction provides a more accurate estimate of nuclear radii than α-particle scattering. This is because electrons interact with nuclei only through the electromagnetic interaction, and this is much better understood than the strong interaction which is involved in α-particle scattering.

Fig. 3.4
Angular distribution of
420 MeV electrons
scattered by $^{16}_{8}O$

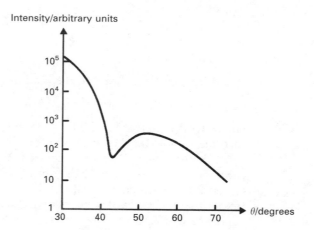

Fig. 3.4 shows the angular distribution of 420 MeV electrons **from a linear accelerator** scattered by a target consisting of $^{16}_{8}O$ nuclei.

A thorough analysis of the data is extremely difficult, but a reasonable estimate of the nuclear radius can be obtained by treating the electrons as waves which have been diffracted by a spherical obstacle – the nucleus – and applying the same theory as that used to interpret the diffraction of light by a spherical object. On this basis the angular position θ of the first diffraction minimum is given by

$$\sin \theta = \frac{0.61\lambda}{R} \tag{3.3}$$

where λ is the de Broglie wavelength of the electrons and R is the radius of the nucleus concerned

The electrons have very high energies and therefore in order to calculate λ it is necessary to use the relativistic formula

$$E = hc/\lambda \tag{3.4}$$

where

E = electron energy (J)

h = Planck's constant (6.63×10^{-34} J s)

c = speed of light (3.00×10^8 m s^{-1})

Here, $E = 420$ MeV $= 420 \times 10^6 \times 1.60 \times 10^{-19} = 6.72 \times 10^{-11}$ J, and equation [3.4] gives $\lambda = 2.96 \times 10^{-15}$ m. From Fig. 3.4, $\theta = 43°$ and it follows from equation [3.3] that $R = 2.65 \times 10^{-15}$ m.

Notes (i)   In order to obtain an appreciable amount of diffraction, λ must be comparable to the nuclear diameter (i.e. very small). It follows from equation [3.4] that electrons with very high energies are needed – hence the linear accelerator.

(ii)   When light is diffracted by objects which have sharp edges the diffraction minima have zero intensity. The minimum in Fig. 3.4 is not of zero intensity, indicating that **the nucleus does not have a well defined edge.**

(iii)   Because nuclei do not have well-defined edges, the various methods of estimating R give slightly different values. $\boldsymbol{R = r_0 A^{1/3}}$ **always applies**, but values of r_0 vary from 1.2×10^{-15} m to 1.5×10^{-15} m. Fig. 3.5 shows a plot of R against $A^{1/3}$ obtained from neutron scattering experiments.

Fig. 3.5
Plot of R against $A^{1/3}$ obtained by scattering 83 MeV neutrons ($r_0 = 1.37 \times 10^{-15}$ m)

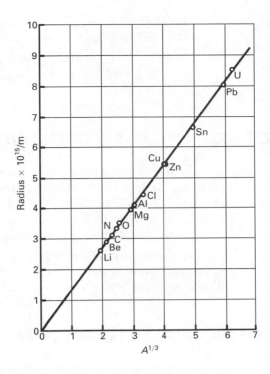

QUESTIONS 3A

1. On the basis of R (in metres) $= 1.4 \times 10^{-15} A^{1/3}$, calculate the radius of **(a)** a gold ($^{197}_{79}$Au) nucleus, **(b)** an α-particle. **(c)** Assuming that the gold nucleus remains at rest throughout, find the initial kinetic energy that an α-particle must have if it is to come just into contact with a gold nucleus **(i)** in joules, **(ii)** in MeV. ($e = 1.6 \times 10^{-19}$ C, $1/4\pi\varepsilon_0 = 9.0 \times 10^9$ m F^{-1}.)

3.5 NUCLEAR DENSITY

Protons and neutrons have approximately equal masses. If the average mass of a nucleon is m, the mass of a nucleus of mass number (nucleon number) A is given by

Mass of nucleus $= Am$

If the nucleus is taken to be a sphere of radius R,

Volume of nucleus $= \frac{4}{3}\pi R^3$

$\qquad\qquad\qquad = \frac{4}{3}\pi r_0^3 A \quad$ (by equation [3.2])

The density, ρ, is therefore given by

$$\rho = \frac{Am}{\frac{4}{3}\pi r_0^3 A}$$

i.e. $\qquad \rho = \dfrac{3m}{4\pi r_0^3}$ $\qquad\qquad\qquad\qquad\qquad\qquad$ [3.5]

Since m and r_0 are constants it follows from equation [3.5] that

All nuclei have the same density

Notes   (i)   We show in Note (ii) of section 3.7 that this is because the force which binds the nucleons inside a nucleus is <u>short-range</u> force.

(ii)   With $r_0 = 1.2 \times 10^{-15}$ m and $m = 1.7 \times 10^{-27}$ kg, equation [3.5] gives $\rho = 2.3 \times 10^{17}$ kg m^{-3}. This is enormous – over 10^{14} times the density of water.

3.6 THE LIQUID-DROP MODEL OF THE NUCLEUS

The fact that all nuclei have the same density suggests that a nucleus can be likened to a drop of liquid, because the density of a drop of water, for example, is the same no matter how big the drop. The **liquid-drop model** of the nucleus has been found useful in accounting for nuclear fission (section 5.1) and for the variation in binding energy with mass number (section 3.8).

The success of the model is associated with the fact that the forces which hold the nucleons together in a nucleus and the forces which hold the molecules together in a liquid drop are both <u>short-range forces</u> – see section 3.7.

3.7 THE STRONG INTERACTION (NUCLEAR FORCE)

The force which holds nucleons together inside a nucleus cannot be the gravitational force, for this is far too weak to overcome the electrostatic repulsion between the protons (see Note (i)). Nor can it be the electrostatic force, for this has no effect on neutrons, and between protons it is actually <u>repulsive</u>! It is supposed, therefore, that nucleons are held together by a force known as **the strong interaction** (sometimes called **the nuclear force**).

Investigations of the interactions between two protons or two neutrons, or between a neutron and a proton, indicate that

> **The strong interaction is charge independent,** i.e. at any given separation the (nuclear) force between two neutrons is the same as that between two protons or between a proton and a neutron.

We can also infer that

> **The strong interaction is a short-range force.**

If it were not, the scattering of α-particles which have not penetrated the nucleus could not be accounted for as being due solely to electrostatic forces (see section 3.3).

The α-scattering results indicate that the strong interaction must be negligible at nucleon–nucleon separations which are not much greater than about 2×10^{-15} m. It must be repulsive at separations of less than about 1.0×10^{-15} m, otherwise nuclei would collapse in on themselves. These and other considerations suggest that the strong interaction between two nucleons probably varies as shown in Fig. 3.6(a). The corresponding variation in the potential energy of the nucleons is shown in Fig. 3.6(b).

Notes   (i)   The electrostatic repulsion F_E and the gravitational attraction F_G between two protons whose separation is r are given by

$$F_E = \frac{1}{4\pi\varepsilon_0} \cdot \frac{e^2}{r^2} \quad \text{and} \quad F_G = G\frac{m_p^2}{r^2}$$

Fig. 3.6
(a) Strong interaction
between two nucleons as
a function of their
separation (b) Potential
energy of two nucleons
as a function of their
separation

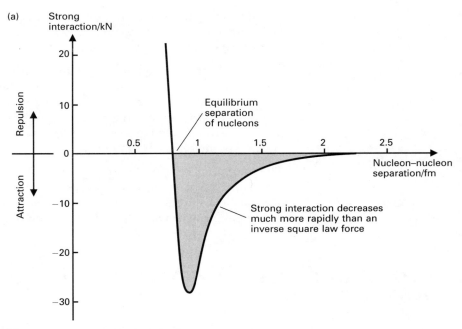

(a)

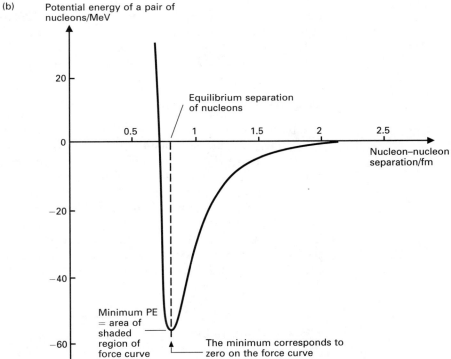

(b)

where m_p is the mass of the proton and G is the gravitational constant. This leads to $F_E/F_G = 1.2 \times 10^{36}$, for all r, i.e. the Coulomb repulsion between two protons is over 10^{36} times as strong as the gravitational attraction between them regardless of their separation.

(ii)   The idea that the strong interaction decreases very rapidly with distance implies that **the nucleons in a nucleus interact only with their nearest neighbours**. This, in turn, implies that **the distance between adjacent nucleons is the same no matter what the size of the nucleus**. (Because each nucleon takes up a position which is determined solely by the forces

(iii) The force between two <u>neutrons</u> is due entirely to the strong interaction and therefore varies as shown in Fig. 3.6(a). The (total) force between two <u>protons</u> includes the contribution of Coulomb repulsion and is therefore not quite the same as shown in Fig. 3.6(a). The difference is significant only where the strong interaction is very close to zero, and cannot be shown on a diagram of the scale of Fig. 3.6(a).

(iv) The minimum PE (Fig. 3.6(b)) is equal to the shaded area of the force–separation curve because this represents the work that would have to be done against the (attractive) strong interaction to allow two nucleons at their equilibrium separation to become infinitely far apart (with PE = 0).

3.8 NUCLEAR STABILITY AND THE LIQUID-DROP MODEL

Energy has to be supplied to a drop of liquid in order to break it up into its component molecules. The energy required is the latent heat of the drop and is proportional to its mass, i.e. to the number of molecules in it. Likewise, energy has to be given to a nucleus to split it up into its component nucleons (see section 2.3). On this basis the binding energy of a nucleus of mass number A should be given by

$$\text{Binding energy} = aA$$

where a is some positive constant.

Nuclei of low mass number have low values of binding energy per nucleon (see Fig. 2.1). This can be explained on the basis of the liquid-drop model as being analogous to a surface tension effect. Nucleons in the interior are attracted from all sides, whereas those in the surface are attracted only from within (Fig. 3.7). The surface nucleons are therefore easier to remove and so have smaller binding energies than those in the interior. In small nuclei, surface nucleons form a greater proportion of the total than they do in larger nuclei. (The <u>surface area</u> of a sphere increases more slowly than its <u>volume</u> does with increasing radius.) It follows that the binding energy is reduced by an amount which is proportional to the surface area, i.e. to R^2, and therefore by equation [3.2], to $A^{2/3}$. An improved expression for binding energy is therefore

$$\text{Binding energy} = aA - bA^{2/3}$$

where b is a positive constant.

Fig. 3.7
To show decreased binding energy of a surface nucleon

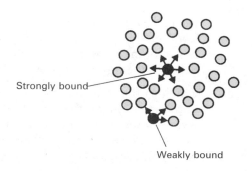

Strongly bound

Weakly bound

Very large nuclei have lower values of binding energy per nucleon than do those of intermediate mass number. This is due to electrostatic repulsion between the protons. These repulsive forces are long-range forces, and therefore every proton is repelled by every other – not just by nearest neighbours. By assuming that the nucleus behaves like a charged liquid drop with charges distributed evenly throughout its volume, it is possible to show that this Coulomb repulsion decreases the binding energy by an amount which is proportional to Z^2/R, i.e. to $Z^2/A^{1/3}$. Therefore

$$\text{Binding energy} = aA - bA^{2/3} - cZ^2/A^{1/3} \qquad [3.6]$$

where c is a positive constant.

Bearing in mind that high binding energy implies high stability, equation [3..6] would lead us to expect that nuclei should not contain protons (for this would make the third term zero) but this, of course, is not so. The stability is decreased by the kinetic energy of the nucleons. It can be shown, but only by using quantum mechanics, that this reduces the binding energy by $d(N - Z)^2/A$ where d is a positive constant. Therefore

$$\text{Binding energy} = aA - bA^{2/3} - cZ^2/A^{1/3} - d(N - Z)^2/A \qquad [3.7]$$

Volume term	Surface term	Electrostatic term	Symmetry term
	Important compared with aA only for small A	Increasingly important as Z increases	Hence the tendency for $N = Z$

The binding energy per nucleon is obtained by dividing the various terms in equation [3.7] by A to give

$$\text{Binding energy per nucleon} = a - bA^{-1/3} - cZ^2/A^{4/3} - d(N - Z)^2/A^2$$

which can be made to fit the observed results (Fig. 2.1) very closely by putting $a = 15.7\,\text{MeV}$, $b = 17.8\,\text{MeV}$, $c = 0.71\,\text{MeV}$ and $d = 23.7\,\text{MeV}$. Fig. 3.8 illustrates the way each of the correction terms depends on mass number, whilst

Fig. 3.8
The contributions of the various correction terms to binding energy per nucleon

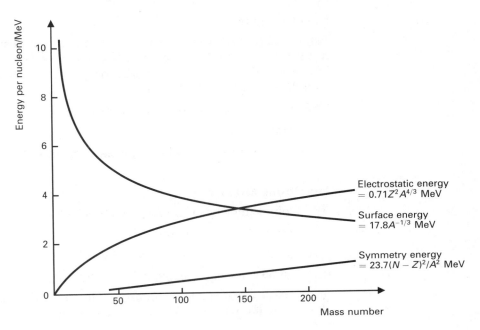

Fig. 3.9 shows the effects of adding successive terms. The bottom curve, which includes all the terms, is indistinguishable from the curve obtained by experiment on the scale used here.

Fig. 3.9
The effects of adding successive terms to the binding energy per nucleon expression

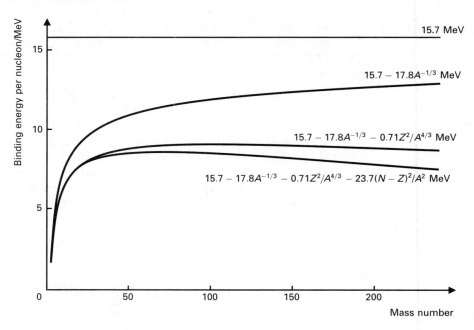

CONSOLIDATION

α-Particle Scattering

Some α-particles are scattered through <u>large</u> angles – suggesting that an atom has a positively charged core (the nucleus).

<u>Only a few</u> α-particles are scattered through large angles – suggesting that the nucleus is very much smaller than the atom as a whole.

Nuclear Radius

Estimated on the basis of α-particle scattering or electron diffraction. Electron diffraction is preferable because it does not involve the (poorly understood) strong interaction. The electrons need high energy so that their wavelength is comparable with the diameter of the nucleus.

The radius of a nucleus of mass number (nucleon number) A is given by

$$R = r_0 A^{1/3}$$

It follows that **all nuclei have the same density**.

The Strong Interaction

(i)   binds nucleons together inside nuclei,

(ii)   is charge independent,

(iii)   is a short-range force and therefore

(iv)   allows nuclei to be likened to liquid drops,

(v)   means that nucleons interact only with their nearest neighbours which explains why all nuclei have the same density.

QUESTIONS ON CHAPTER 3

1. **(a)** Geiger and Marsden aimed α-particles of mass 6.7×10^{-27} kg and speed 2.0×10^7 m s^{-1} at gold nuclei for which the number of protons is 79.
   Calculate the closest distance of approach for a head-on collision, assuming that the gold nucleus remains stationary.
   (Charge of a proton $= 1.6 \times 10^{-19}$ C, permittivity of free space, $\varepsilon_0 = 8.9 \times 10^{-12}$ F m^{-1}.)
   (b) Describe how electron diffraction is used in determining nuclear sizes. Explain why this method is used in preference to α-particle scattering experiments. [J]

2. **(a)** Briefly describe the principles involved in estimating the size of a nucleus from experiments in which alpha-particles are scattered by target materials such as gold foil.
   (b) Calculate the initial kinetic energy, in MeV, required by an alpha-particle if it is to reach the closest distance of approach when travelling head-on towards the nucleus of a gold atom ($^{197}_{79}$Au). The radius of the gold nucleus may be taken to be 7.6 fm, and that of the alpha-particle to be 2.0 fm.
   (c) What are the values of the energy, in MeV, of naturally occurring alpha-particles from radioactive sources, such as those used in the original scattering experiments? Suggest a method by which alpha-particles could be made more penetrating. Also explain why electrons are generally preferred to alpha-particles when probing the nucleus.
   ($e = 1.6 \times 10^{-19}$ C, $\varepsilon_0 = 8.9 \times 10^{-12}$ F m^{-1}.) [J, '91]

3. **(a)** The radius, R, of an atomic nucleus can be determined by bombardment with high energy electrons. Name an apparatus suitable for producing electrons of the required energy, and explain why this energy has to be high.
   State the equation which relates R to nucleon number, A. If the radius of the nucleus of $^{93}_{37}$Rb is 5.89×10^{-15} m, calculate the radius of the nucleus of $^{143}_{55}$Cs.
   (b) When $^{235}_{92}$U is bombarded with neutrons, $^{93}_{37}$Rb and $^{143}_{55}$Cs are possible fission products. Assuming that these two nuclei are initially in contact, calculate the total

kinetic energy they would acquire due to Coulomb repulsion.
(Charge of a proton $= 1.6 \times 10^{-19}$ C, permittivity of free space, $\varepsilon_0 = 8.9 \times 10^{-12}$ F m^{-1}.) [J]

4. The table gives a series of values of R, the radius of the nucleus, as determined from electron scattering experiments, for nuclei of nucleon number A.

Element	A	R/10^{-15} m
Aluminium	27	3.60
Vanadium	51	4.50
Strontium	88	5.34
Caesium	133	6.15
Bismuth	209	7.13

   (a) **(i)** State the relationship between R and A and use the data given in the table to plot a suitable straight line graph to verify this relationship. Use your graph to calculate the density of nuclear matter, if the mass of a nucleon is 1.7×10^{-27} kg.
   (ii) What properties of the forces between nucleons can be deduced from this relationship between R and A?
   (b) Estimate the energy of the electrons used to obtain the value of R for aluminium shown in the table, assuming that at this energy
   $$\frac{\text{Energy}}{\text{Momentum}} = \text{The speed of light.}$$
   (The Planck constant $= 6.6 \times 10^{-34}$ J s, speed of light in vacuo $= 3.0 \times 10^8$ m s^{-1}.) [J]

5. **(a)** Explain why electron diffraction is used to determine nuclear size.
   (i) Estimate the momentum the electrons must be given.
   (ii) Name an apparatus capable of producing electrons having appropriate speeds. (Planck's constant $h = 6.6 \times 10^{-34}$ J s.)
   (b) What are the main characteristics of the interaction between nucleons in the nucleus? Show how they account for the relationship $R = r_0 A^{1/3}$, where
   R = nuclear radius
   A = atomic mass number
   r_0 = a constant. [J]

6.

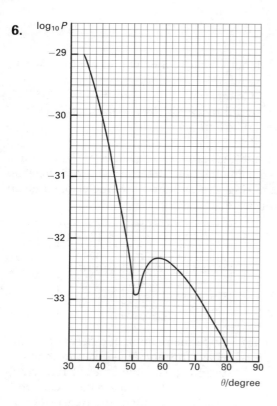

θ/degree

The graph shows the results of electron scattering experiments for ^{12}C, in which electrons of energy 420 MeV are used. P is a measure of the number of electrons scattered at angle θ to the incident beam.

(a) Show that the de Broglie wavelength, λ, of the incident electrons is 2.96×10^{-15} m, assuming that the relation

$$\frac{\text{Energy}}{\text{Momentum}} = c, \text{ the speed of light,}$$

applies at this energy.
(The charge of an electron $= -1.60 \times 10^{-19}$ C, speed of light $= 3.00 \times 10^8 \text{ms}^{-1}$, the Planck constant $= 6.63 \times 10^{-34}$ J s.)

(b) Give *two* reasons why high energy electrons are used in determining nuclear size. What information do such measurements give about nuclear density and the average separation of particles in the nucleus?

(c) In diffraction theory, the first minimum of the scattered intensity produced by a spherical object of radius R occurs at an angle θ to the direct beam, where θ is given by

$$\sin \theta = \frac{0.61 \lambda}{R}$$

Use the experimental data to calculate the radius of the ^{12}C nucleus.
Use your result to calculate the radius of the ^{16}O nucleus.   [J]

7. (a) Sketch a graph to show how the potential energy of two neutrons varies with their separation for values of separation comparable with the sizes of nuclei. Show approximate values on the separation axis of the graph, mark the equilibrium separation of the neutrons, and indicate the regions in which there is **(i)** a net attractive force, **(ii)** a net repulsive force, between the neutrons.

(b) High energy electrons are diffracted by nuclei in a similar way to the diffraction of light through a circular aperture, giving a pattern of circular rings.

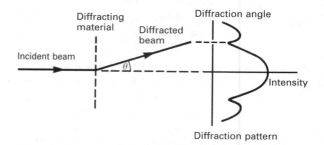

It may be shown that the first minimum of the diffraction pattern is observed at angle θ, where

$$\sin \theta = \frac{1.22 \lambda}{d}$$

λ = wavelength of incident beam
d = diameter of circular aperture.

If the first minimum occurs at $\theta = 44°$ when electrons of energy 420 MeV are incident on oxygen nuclei, calculate

(i) the momentum of the incident electrons, assuming that they obey the relationship Energy = Momentum × Speed of light in a vacuum,

(ii) the wavelength of the incident electron waves,

(iii) the radius of the oxygen nucleus.
What property of the nucleons in nuclei can be deduced by studying the results from experiments such as this?

(c) The volume of a $^{28}_{14}\text{Si}$ nucleus may be taken to be 2.5×10^{-43} m³. Use this value to calculate

(i) the radius of the $^{28}_{14}\text{Si}$ nucleus,

(ii) the volume of a $^{120}_{50}\text{Sn}$ nucleus.
State any assumptions you make in your calculation.
($e = 1.60 \times 10^{-19}$ C, $c = 3.00 \times 10^8 \text{ m s}^{-1}$, $h = 6.63 \times 10^{-34}$ J s.)   [J, '92]

8. (a) Describe how high energy electron scattering has been used to determine nuclear sizes. How do nuclear sizes depend on the mass number, A?

(b) Describe the liquid drop model of the nucleus, showing how the model helps in understanding the variation of nuclear binding energy with A and the charge number, Z.

9. (a) (i) State the order of magnitude of the radius of a nucleus.

(ii) By considering a nucleus of nucleon number A and radius R, show that the density of a nucleus is independent of its nucleon number. With what property of the forces between nucleons is this result consistent?

(b) Sketch a graph showing how the force varies with the separation for two protons at separations comparable with the sizes of nuclei. Show approximate scales ont he axes of the graph, and indicate clearly the regions where the net force is attractive and where it is repulsive.

(c) 'In order to investigate the size of a nucleus by electron diffraction techniques, electrons should be accelerated across a p.d. of about 1000 MV.'

Give a quantitative justification for this statement, using the following data.

(The Planck constant, $h = 6.6 \times 10^{-34}$ Js, speed of electromagnetic radiation in vacuo, $c = 3.0 \times 10^8 \,\mathrm{m\,s^{-1}}$, charge of an electron, $e = 1.6 \times 10^{-19}$ C.)

(It may be assumed that, at the required electron energy, $\dfrac{\text{Energy}}{\text{Momentum}} = c$.)   [J]

10. (a) What is meant by the binding energy of a nucleus? Sketch a graph of binding energy per nucleon against nucleon number for the naturally occurring nuclides, indicating approximate scales on the axes of your graph.

(b) One form of the binding energy equation of the liquid-drop model for a nucleus of nucleon number A, proton number Z and neutron number N is given below, where a, b, c and d are constants.

$$\text{Binding energy} = aA - bA^{2/3} - \frac{cZ^2}{A^{1/3}} - \frac{d(N-Z)^2}{A}$$

(i) State the origins of the terms in the equation containing b, c and d.

(ii) Explain how the equation is consistent with the general shape of the graph in (a), indicating which terms are more important for light nuclei and which for heavy nuclei.

(c) What are the main characteristics of the strong forces between nucleons in the nucleus?   [J]

11. (a) The graph indicates how the potential energy of a pair of neutrons in a nucleus may be considered to vary with their separation.

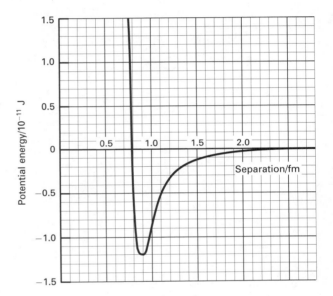

(i) If the neutrons were stationary in the nucleus, at what separation ought they to be in equilibrium according to the graph? Explain your answer.

(ii) Estimate the energy, in MeV, which would have to be supplied to a pair of stationary neutrons in order to separate them completely, starting from their equilibrium position.

(iii) What value of binding energy per neutron is suggested by your answer to (ii)? How does this compare with the actual value of binding energy per nucleon for most of the nuclei where each nucleon is surrounded by several near neighbours? Suggest an explanation for the discrepancy between these values.

(iv) What physical quantity is represented by the gradient of the graph? Explain the significance of the different sign and magnitude of the gradients on either side of the equilibrium position.

(v) Use the graph to estimate the magnitude of the nuclear force when the separation is 1.25 fm.

(b) (i) State *two* reasons why high energy electrons are suitable for probing the nucleus by diffraction techniques.

(ii) When electron waves of wavelength λ are incident on nuclei of radius R, the first minimum of the scattered intensity occurs at angle θ from the direct beam, where

$$\sin \theta = \frac{0.61\,\lambda}{R}$$

Assuming the electrons obey the relationship

Energy = Momentum × Speed of light in a vacuum

Show that

$$\sin \theta = \frac{0.61hc}{Er_0\,A^{1/3}}$$

where E is the energy of the incident electrons, A is the nucleon number of the target nuclei and r_0 has its usual meaning.

(iii) Use the results from (ii) to decide the maximum and minimum energy values, in MeV, which would be suitable for investigating the first minimum of the $^{12}_{6}C$ nucleus by this technique. The value of the constant r_0 in the equation $R = r_0\,A^{1/3}$ may be taken to be 1.3 fm.   [J, '94]

4

DECAY MECHANISMS

4.1 EXCITED STATES OF NUCLEI

When a nucleus decays the daughter nucleus may be produced in its ground state or in one of a small number of excited states. If it is in one of the excited states, it decays to its ground state, either directly or indirectly, by emitting a γ-ray (or rays). **The γ-ray(s) is created at the instant of decay** – it has not existed previously.

Consider the decay of $^{228}_{90}$Th to $^{224}_{88}$Ra by α-emission (Fig. 4.1). α-particles of five different energies can be produced (though not all with the same probability). The emission of an α-particle with 5.140 MeV, 5.176 MeV, 5.211 MeV or 5.342 MeV leaves the daughter in an excited state. It then decays to its ground state by emitting a single γ-ray (corresponding to the transitions D or E) or two γ-rays (transitions A and E, B and E or C and E).

Fig. 4.1
γ-emission from excited states of $^{224}_{88}$Ra

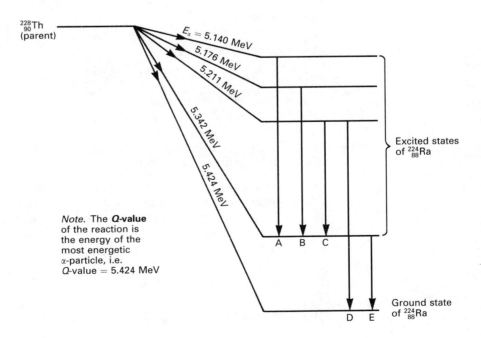

Reference to Table 4.1 shows that the γ-ray energies are approximately equal to the difference in energy of the α-particles associated with the energy levels concerned. The agreement is not exact because no account has been taken of the recoil energy of the $^{224}_{88}$Ra.

Note β-emitters also produce γ-rays with a small number of discrete energies.

Table 4.1
Comparison of α-particle
and γ-ray energies

Transition	γ-ray energy/MeV	Energy difference of the α-particles/MeV
A	0.205	$5.342 - 5.140 = 0.202$
B	0.167	$5.342 - 5.176 = 0.166$
C	0.132	$5.342 - 5.211 = 0.131$
D	0.216	$5.424 - 5.211 = 0.213$
E	0.085	$5.424 - 5.342 = 0.082$

4.2 BETA DECAY

There are three types of β-decay: negative β-decay (electron emission), positive β-decay (positron emission) and electron capture.

Electron emission and positron emission have been discussed previously (sections 1.5 and 2.6).

In **electron capture** a nuclear proton captures an electron from the electron cloud surrounding the nucleus and becomes a neutron. The process is normally accompanied by the emission of an X-ray, created when an outer electron falls into the vacancy left behind by the electron that has been captured.

β-decay also involves the emission of a neutrino or an antineutrino (see section 4.3) according to:

β⁻-decay $\quad {}_Z^A X \rightarrow {}_{Z+1}^A Y + {}_{-1}^0 e + {}_0^0 \bar{\nu}$ [4.1]
Electron   Antineutrino

β⁺-decay $\quad {}_Z^A X \rightarrow {}_{Z-1}^A Y + {}_1^0 e + {}_0^0 \nu$ [4.2]
Positron   Neutrino

Electron capture $\quad {}_Z^A X + {}_{-1}^0 e \rightarrow {}_{Z-1}^A Y + {}_0^0 \nu$ [4.3]
Captured   Neutrino
electron

Note There are actually three kinds of neutrino and antineutrino (see section 6.8); those referred to here are electron neutrinos.

The electrons, positrons, neutrinos and antineutrinos do not exist within nuclei prior to emission. They are created at the instant of decay when a proton becomes a neutron, or a neutron becomes a proton. The three processes are:

β⁻-decay $\quad {}_0^1 n \rightarrow {}_1^1 p + {}_{-1}^0 e + {}_0^0 \bar{\nu}$ [4.4]

β⁺-decay $\quad {}_1^1 p \rightarrow {}_0^1 n + {}_1^0 e + {}_0^0 \nu$ [4.5]

Electron capture $\quad {}_1^1 p + {}_{-1}^0 e \rightarrow {}_0^1 n + {}_0^0 \nu$ [4.6]

The first of these involves a decrease in mass; it is therefore energy favourable and can also occur outside a nucleus – free neutrons decay with a mean lifetime* of about 15 minutes. The other two processes ([4.5] and [4.6]) involve increases in mass and do not occur spontaneously with free protons. When the nucleus as a whole is taken into account ([4.2] and [4.3]) there is an overall decrease in mass.

*This should not be confused with half-life – see Note (ii) on p. 95.

4.3 CONTINUOUS ENERGY SPECTRUM OF β-DECAY

β-particles (unlike α-particles and γ-rays) are emitted with a <u>continuous</u> range of energies

A typical energy spectrum is shown in Fig. 4.2. This continuous distribution of β-particle energies would not be possible were it not for the fact that neutrinos (or antineutrinos) are also emitted, for the process would not be able to satisfy <u>both</u> the law of conservation of momentum and the law of conservation of energy. If the decay resulted only in a β-particle and the daughter nucleus, then in order to conserve momentum, the two particles would have to move in <u>opposite</u> directions with speeds which were in the inverse ratio of their masses. Since their combined energy has to equal that of the decay process (**the Q-value**), fulfilment of this condition would require that all the β-particles would have the same energy (as would all the daughter nuclei).

Fig. 4.2
Typical energy
distribution of β-particles

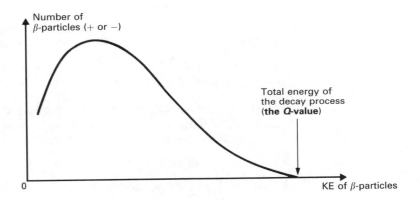

The emission of a third particle (a neutrino or an antineutrino) allows the β-particle to move in any direction relative to the daughter nucleus and with any energy up to the maximum available, for the neutrino simply moves in such a direction and with such energy (see Note (i)) that both momentum and energy are conserved (Fig. 4.3).

It might be thought that the continuous range of β-particle energies could be explained by there being γ-rays emitted along with the β-particles. This cannot be the explanation – γ-rays are not <u>always</u> emitted and they never have a <u>continuous</u> range of energies.

Fig. 4.3
Momentum conservation
in β-decay

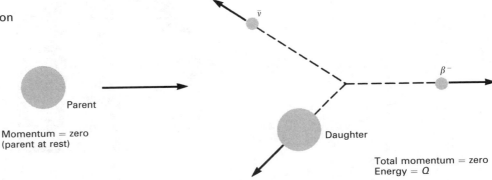

Notes (i) Although all neutrinos (and antineutrinos) move at the speed of light, they do not all have the same energy or momentum. (The same is true, of course, of photons.) When the energy of the neutrino is zero the total disintegration energy, **the Q-value**, is shared between the β-particle and the recoiling daughter nucleus. Since the nucleus is so much more massive than the β-particle, it has only about 0.001% of the available energy and, to a very good approximation, the β-particle energy is the Q-value.

(ii) Tables listing β-particle energies give the <u>maximum</u> values for the nuclides concerned – **the Q-values**.

4.4 THE NEUTRINO

The existence of the neutrino was first postulated in 1930 by Wolfgang Pauli in order to explain observations concerning β-decay (section 4.3). The neutrino has the following properties.

Zero charge otherwise charge would not be conserved in β-decay.

Zero rest mass (or very nearly so) because, within experimental error, Q-values calculated on the basis of it being zero are equal to the maximum β-particle energy plus the energy of the recoiling nucleus. Also, theoretical predictions of the shape of the high-energy tail of the β-spectrum involve the mass of the neutrino. The observed shapes indicate that the mass is less than 1/2000 of the electron mass.

Velocity = Speed of light on the assumption of zero rest mass.

Spin = $\frac{1}{2}$ ★ so that angular momentum is conserved.

Virtually no interaction with matter. The mean free path in a block of lead has been calculated to be several hundreds of light-years. Neutrinos interact by means of **the weak interaction**.

Wolfgang Pauli (1900–58)

★See section 6.8.

The **antineutrino** has the same five properties. The two particles are distinguished in that their spins (see section 6.8) are in opposite directions when compared with their directions of motion (Fig. 4.4).

Fig. 4.4
Distinction between neutrino and antineutrino

Neutrino

Antineutrino

Because neutrinos and antineutrinos are uncharged, they can be detected only by indirect means. This was first achieved in 1956 by Cowen and Reines. The method was to detect the products of a reaction that had been initiated by an antineutrino. They placed a large tank containing an organic liquid close to a nuclear (fission) reactor. The reactor provided a large flux of antineutrinos, produced by negative β-decay of the fission products. The antineutrinos were absorbed by protons in the liquid according to:

$$_0^0\bar{\nu} + {}_1^1\text{p} \rightarrow {}_0^1\text{n} + {}_1^0\text{e}$$

The simultaneous creation of a neutron (${}_0^1\text{n}$) and a positron (${}_1^0\text{e}$) was taken as evidence that the reaction had taken place, and therefore, that the antineutrino did in fact exist. (Note, the neutron and the positron were also detected by indirect means.)

Note The energy, E, and momentum, p, of a neutrino are related by the relativistic formula for particles of zero rest mass,

$$E = pc \qquad \left(\begin{array}{c}\text{For a particle of}\\ \text{zero rest mass}\end{array}\right)^\star$$

Both E and p can have any value from zero up to the maximum available.

4.5 BARRIER PENETRATION (TUNNELLING) IN α-DECAY

α-particle emission is quite distinct from β and γ-emission in that **α-particles exist within nuclei long before they are emitted**, whereas β-particles and γ-rays are created at the instant of decay. The protons and neutrons inside a nucleus are thought to move around inside it at very high speeds. When two protons and two neutrons come into particularly close proximity they may stay together as a **cluster** for some considerable time, because to do so is energetically favourable as it allows an amount of energy equal to the binding energy of an α-particle (≈ 28 MeV) to be released.

An α-particle <u>inside</u> a nucleus feels a strongly attractive force due to the strong interaction. <u>Outside</u>, however, it feels a repulsive force due to Coulomb repulsion. As a result, the potential energy of an α-particle varies with distance from the centre of its nucleus as shown in Fig. 4.5. The horizontal line represents the total energy (PE + KE) of a typical α-particle. This is always less than 10 MeV (and on average is only about 6 MeV). According to classical physics it should be impossible for such an α-particle ever to leave the nucleus for if it were to cross the potential barrier, its potential energy would need to exceed 25 MeV.

$\star E = pc$ is a good approximation for <u>any</u> particle providing its speed is close to the speed of light.

Fig. 4.5
Variation of α-particle PE
as a function of distance
from centre of nucleus

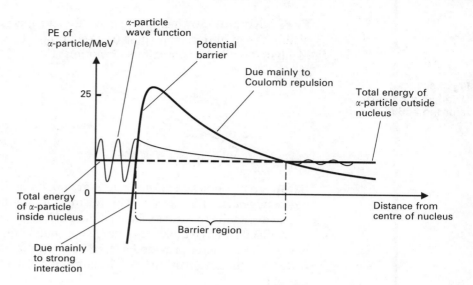

According to quantum mechanics, the α-particle can be thought of as a wave. The amplitude of this wave is large inside the nucleus wherever the 'height' of the barrier is less than the energy of the α-particle. Although it decreases quite drastically within the barrier region, it does not become zero and so continues to exist outside the barrier, i.e. outside the nucleus. The probability of the α-particle being at any particular point is proportional to the square of the amplitude of the wave at that point. There is therefore a small but finite probability of finding the α-particle outside the nucleus. The α-particle is said to have **tunnelled** through the barrier.

An α-particle moving around inside a nucleus repeatedly comes up against the potential barrier. The probability of it tunnelling through on any one occasion is extremely low. In the case of uranium 238, for example, it has been calculated that an α-particle makes 10^{38} attempts before it is successful!

4.6 RELATIONSHIP BETWEEN α-PARTICLE ENERGY AND HALF-LIFE

α-emitters with long half-lives emit α-particles with lower energy than those with short half-lives.

A plot of log (half-life) against log (α-particle energy) has the form shown in Fig. 4.6, i.e. α-emitters obey the relationship

$$\log T_{1/2} = A - B \log E_\alpha$$

where A and B are positive constants.

Fig. 4.6
Relationship between
half-life and α-particle
energy

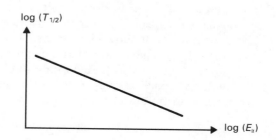

There is considerable scatter about the line when all known α-emitters are included. There is less scatter when the three radioactive series (section 1.12) are treated separately – the results fall on three closely-spaced lines of equal slope. Plots restricted to single values of Z are extremely smooth.

The relationship between α-particle energy and half-life is consistent with the theory of α-particle tunnelling. The probability of an α-particle being able to tunnel through the barrier depends on the thickness of the barrier that it is required to penetrate. The higher the total energy of the α-particle, the narrower the barrier it is presented with (Fig. 4.7) and therefore the greater the possibility of escape and the lower the half-life.

Fig. 4.7
Relationship between α-particle energy and barrier thickness

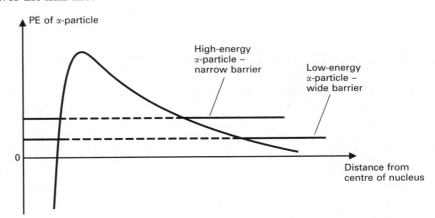

CONSOLIDATION

γ-ray emission

Any particular nuclide emits γ-rays with a small number of discrete energies. (This provides evidence for the existence of nuclear energy levels.)

γ-rays are emitted by daughter nuclei which have been produced in an excited state following α or β-emission.

γ-rays are created at the instant of decay.

β^--emission

$$^{A}_{Z}\text{X} \rightarrow ^{A}_{Z+1}\text{Y} + ^{0}_{-1}\text{e} + ^{0}_{0}\overline{\nu} \qquad \left(^{1}_{0}\text{n} \rightarrow ^{1}_{1}\text{p} + ^{0}_{-1}\text{e} + ^{0}_{0}\overline{\nu}\right)$$

β^+-emission

$$^{A}_{Z}\text{X} \rightarrow ^{A}_{Z-1}\text{Y} + ^{0}_{1}\text{e} + ^{0}_{0}\nu \qquad \left(^{1}_{1}\text{p} \rightarrow ^{1}_{0}\text{n} + ^{0}_{1}\text{e} + ^{0}_{0}\nu\right)$$

β-particles are created at the instant of decay. β-particles have a continuous range of energies. (This can occur without violation of either conservation of momentum or conservation of energy because neutrinos or antineutrinos are emitted as well.)

Electron capture

$$^{A}_{Z}\text{X} + ^{0}_{-1}\text{e} \rightarrow ^{A}_{Z-1}\text{Y} + ^{0}_{0}\nu \qquad \left(^{1}_{1}\text{p} + ^{0}_{-1}\text{e} \rightarrow ^{1}_{0}\text{n} + ^{0}_{0}\nu\right)$$

Electron capture is usually accompanied by X-ray emission.

α-emission

α-particles exist inside nuclei <u>before</u> they are emitted.

Classically, α-particles do not have sufficient energy to escape from nuclei. According to quantum mechanics, they escape by 'tunnelling' through the potential barrier.

α-emitters with long half-lives produce low-energy α-particles. (This is consistent with the theory of tunnelling.)

$$\log T_{1/2} = A - B \log E_{\alpha}$$

QUESTIONS ON CHAPTER 4

1. When ^{219}Rn decays it emits α-particles of two distinct energies, 6.8 MeV and 6.4 MeV. The value of Q, the energy released, calculated from the masses involved, is 6.9 MeV.
  (a) Explain why the more energetic α-particle has less energy than the calculated Q-value.
  (b) Explain why the α-particles have two distinct energies. State what radiation you would look for to confirm your explanation, and estimate its energy.   [J]

2. The decay scheme for $^{27}_{12}$Mg is shown in the figure.

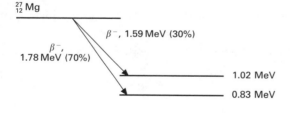

  (a) (i) Identify the daughter nuclide.
  (ii) Briefly describe the decay sequence represented by the figure.
  (b) (i) Write down an equation to represent the initial decay of a $^{27}_{12}$Mg nucleus.
  (ii) Calculate the Q-value, in MeV, for the reaction represented by the equation in (b)(i). Explain how you account for the electron masses.
  (iii) Name another form of radiation to be expected in this decay scheme in addition to beta radiation, and give *three* possible values for its energy.

  (iv) Calculate the wavelength of the photon with the highest energy that might be expected to occur in this decay.
  (v) By using the decay scheme outlined in the figure, estimate the total energy carried off by the radiations when $^{27}_{12}$Mg decays, explaining how you arrive at your answer. Why might this answer differ slightly from the Q-value in (b)(ii)?
  (c) (i) Explain why it became necessary, in the development of theories about the nucleus, to propose the existence of neutrinos and antineutrinos.
  (ii) State *one* difference between the physical properties of the neutrino and the antineutrino.

Name	Symbol	Atomic mass/u
Sodium	$^{23}_{11}$Na	22.98977
	$^{24}_{11}$Na	23.99097
	$^{25}_{11}$Na	24.98990
Magnesium	$^{24}_{12}$Mg	23.98505
	$^{25}_{12}$Mg	24.98584
	$^{26}_{12}$Mg	25.98259
	$^{27}_{12}$Mg	26.98435
	$^{28}_{12}$Mg	27.98388
Aluminium	$^{25}_{13}$Al	24.99041
	$^{26}_{13}$Al	25.98690
	$^{27}_{13}$Al	26.98154
	$^{28}_{13}$Al	27.98191

$(1\,u = 931.3\,\text{MeV}, \quad e = 1.60 \times 10^{-19}\,\text{C},$
$h = 6.63 \times 10^{-34}\,\text{J s}, \quad c = 3.00 \times 10^{8}\,\text{m s}^{-1}.)$

[J, '91]

3. (a) Describe the properties of the neutrino. How does the postulate of the existence of the neutrino account for the observed features of beta ray energy spectra?

(b) The nuclide ^{37}Ar can decay by electron capture.
(i) Explain what is meant by *electron capture*.
(ii) Calculate the energy released in the decay of ^{37}Ar, using data selected from the table below.
(iii) State how the energy released is carried off following the decay.

In the following table, the mass of an isotope is given for a neutral atom of the substance and is quoted in unified atomic mass units, u. 1 u is equivalent to 931.3 MeV.

Name	Symbol	Charge number Z	Mass number A	Atomic mass M/u
Electron	$e^{\pm}$	± 1	—	0.00055
Proton	p	1	1	1.00728
Neutron	n	0	1	1.00867
Hydrogen	^{1}H	1	1	1.00783
Helium	^{4}He	2	4	4.00260
Chlorine	^{35}Cl	17	35	34.96885
	^{36}Cl	17	36	35.96831
	^{37}Cl	17	37	36.96590
	^{38}Cl	17	38	37.96800
Argon	^{36}Ar	18	36	35.96755
	^{37}Ar	18	37	36.96677
	^{38}Ar	18	38	37.96572
	^{39}Ar	18	39	38.96432
	^{40}Ar	18	40	39.96238
Potassium	^{37}K	19	37	36.9734
	^{38}K	19	38	37.9691
	^{39}K	19	39	38.96371
	^{40}K	19	40	39.96401
	^{41}K	19	41	40.96183

[J]

4. (a) There are *three* different spontaneous processes in which the proton number of an unstable nucleus changes by one. Give equations for each of these processes, state what changes occur within the nucleus and what particles are emitted or absorbed. In each case state whether or not radiation may be emitted, and explain your answer.

(b) State *four* differences between the physical decay mechanisms of alpha and beta-minus emission, apart from the changes they produce in the proton number and the nucleon number.

(c) The nuclide $^{14}_{6}$C decays by beta-minus emission. Explaining each step in your working, calculate
(i) the Q-value, in J, for this reaction,
(ii) the mass of pure $^{14}_{6}$C which would have to undergo decay in order to release 100 MJ of energy.
(Atomic mass of $^{14}_{6}$C = 14.00324 u, and of $^{14}_{7}$N = 14.00307 u. 1 u = 1.661 × 10^{-27} kg, c = 3.00 × 10^8 m s^{-1}, N_A = 6.02 × 10^{23} mol^{-1}.) [J]

5. (a) (i) Sketch a graph showing the potential energy of an α-particle near a heavy nucleus as a function of their separation. State what is meant by the tunnel effect and use your statement to explain how a nucleus can decay by emitting an α-particle.
(ii) Account for the fact that ^{226}Ra emits α-particles of two distinct energies accompanied by γ-radiation. Calculate the wavelength of the γ-radiation if the energies of the α-particles are 7.65 × 10^{-13} J and 7.36 × 10^{-13} J. (Speed of light = 3.00 × 10^8 m s^{-1}, the Planck constant = 6.63 × 10^{-34} J s.)

(b) Sketch a typical kinetic energy spectrum for β^--particles produced in a radioactive decay. Explain the nature of this spectrum, and state how it is characteristic of the decaying nuclide. [J]

6. (a) Sketch a typical kinetic energy spectrum for negative β-particles emitted in a radioactive decay with a Q-value of 200 keV. Explain the nature of this spectrum.

(b) Explain how a nucleus can decay by emitting an α-particle. Use your explanation to account for the observation that isotopes having short half-lives emit higher energy α-particles than isotopes having long half-lives.

(c) When $^{224}_{88}$Ra decays, it emits α-particles of two distinct energies, 5.68 MeV and 5.45 MeV. Explain this, and state what evidence you would look for to confirm your explanation. [J]

7. (a) $^{235}_{92}$U $\xrightarrow[4.6, 4.4, 4.1]{\alpha}$ Th $\xrightarrow[0.3]{\beta}$ Pa $\xrightarrow[5.1, 4.7]{\alpha}$

Ac $\xrightarrow[0.04]{\beta}$ X $\xrightarrow[5.9, 5.7, 5.4]{\alpha}$ Ra

Part of a natural radioactive decay series is shown on the previous page.. The type of particle emitted by each nucleus is indicated above the arrow, while the numbers below are the energies in MeV which are associated with these particles and which are normally given in charts of the decay series.

(i) Give the atomic number and the mass number of nuclide X.

(ii) Explain why there are no β^+-emitters in the part of the series shown.

(iii) Explain the significance of the single energy value given for each β^--emitter.

(iv) Explain why many α-emitters produce α-particles of several different energies.

(v) Which α-emitter of those included would you expect to have the shortest half-life? Give a reason for your answer.

(vi) Name one isotope in the above series which might emit an antineutrino. Justify your answer.

(b) State one method of producing artificial radioactive isotopes. Suggest *one* use of one such isotope. [J]

8. (a)

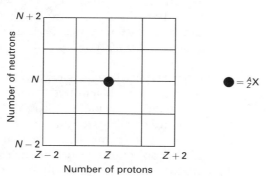

Number of protons

The outline grid above is a neutron–proton diagram showing the region around the nuclide $_Z^A X$, which contains N neutrons. Copy this diagram.

Draw lines on your diagram indicating clearly each of the changes which would be produced if $_Z^A X$ were to decay separately by the following alternative modes:

(i) by α-emission, becoming nuclide R

(ii) by β^--emission, becoming nuclide S

(iii) by electron capture, becoming nuclide T

(iv) by neutron emission, becoming nuclide U

Mark R, S, T and U clearly on your diagram. What are the principal characteristics of a nucleus which is likely to decay by emitting a neutron directly? Give an example of a process which could lead to the production of such a nucleus.

(b) (i) Write down an equation to represent the decay of the nuclide $_Z^A X$ into the nuclide S above.

(ii) Sketch a graph to show the distribution of energies of β-particles when a nuclide decays by β^--emission. Describe and explain the principal features of this graph. [J, '94]

9. (a) (i) Calculate the potential energy of an α-particle in MeV when at a distance of 2.0×10^{-14} m from a $_{82}^{206}$Pb nucleus. (Charge of a proton $= 1.6 \times 10^{-19}$ C, permittivity of free space, $\varepsilon_0 = 8.9 \times 10^{-12}$ F m^{-1}.)

(ii) The graph shows the Coulomb potential barrier for an α-particle near a ^{206}Pb nucleus. The vertical scale, which is linear, has been omitted.

Use the result of your calculation in (i) to determine the vertical scale and hence show that the maximum height of the barrier is about 22 MeV.

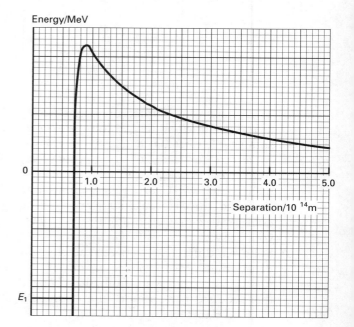

(iii) Use the data below to calculate the binding energy which becomes available if an α-particle cluster is formed from two protons and two neutrons. From this result and the scale of the graph, estimate the energy of the α-particle which the nucleus would emit, if E_1 is the total energy level of two neutrons and two protons bound within the nucleus of $^{210}_{84}$Po.

Indicate briefly how it is possible for the α-particle to escape.

(Mass of a proton = 1.0073 u, mass of a neutron = 1.0087 u, mass of an α-particle = 4.0015 u. 1 u is equivalent to 931 MeV.)

(b) The nucleus $^{203}_{82}$Pb can decay by electron capture. Explain what is meant by this statement.

Give one reason why electromagnetic radiation accompanies electron capture, stating what form it would take.

[J]

10. (a) Draw up a table of $\log_{10}$[half-life] and particle energy for the *α-particle emitters* listed in Table 1. Only list your values of $\log_{10}$[half-life] to two significant figures.

(b) Plot a graph of $\log_{10}$[half-life] (y-axis) against particle energy, in MeV (x-axis).

(c) What conclusions about the connection between half-life and particle energy can be drawn from the graph?

Table 1 The Radioactive Series of $^{238}_{92}$U

Nuclide	Decay mode	Half-life /s	Particle energy/MeV
^{238}U	α	1.4×10^{17}	4.20
^{234}Th	β	2.1×10^6	0.19
^{234}Pa	β	7.1×10^1	2.32
^{234}U	α	7.9×10^{12}	4.77
^{230}Th	α	2.5×10^{12}	4.68
^{226}Ra	α	5.1×10^{10}	4.78
^{222}Rn	α	3.3×10^5	5.49
^{218}Po	α	1.8×10^2	6.00
^{214}Pb	β	1.6×10^3	0.70
^{214}Bi	β	1.2×10^3	3.17
^{214}Po	α	1.6×10^{-4}	7.68
^{210}Pb	β	6.1×10^8	0.02
^{210}Bi	β	4.3×10^5	1.15
^{210}Po	α	1.2×10^7	5.30
^{206}Pb	Stable		

[AEB, '90]

5

FISSION AND FUSION

The fission process has been discussed in section 2.10. The reader should be familiar with this before continuing.

5.1 FISSION AND THE LIQUID-DROP MODEL

Like the nucleons in a nucleus, the molecules in a drop of liquid are held together by short-range forces. Thus the molecules on one side of a drop are not influenced by those on the other side. Similarly, the nucleons on one side of a nucleus feel no effects from those on the other side of the nucleus.

The molecules in the surface of a drop have more energy (i.e. less binding energy) than those in the interior. Suppose that a spherical drop has gained energy by some means and, as a result, has become elongated. It can reduce its energy (and therefore become more stable) by breaking into two smaller but more spherical drops which, because they are more spherical, have a smaller combined surface area than the elongated drop. It is because of the short-range nature of the forces between the molecules that the elongated drop breaks into two rather than reforms into its original spherical shape.

When a $^{235}_{92}$U nucleus absorbs a neutron the resulting $^{236}_{92}$U nucleus is in a highly energetic state on account of the kinetic energy supplied by the neutron and the binding energy released by its absorption. This sets up an oscillation which can distort the nucleus so much that it elongates. It can find a more stable state by forming two approximately spherical lobes separated by a narrow neck (Fig. 5.1). Once the neck has formed the influence of the short-range forces holding the two parts together is very much reduced and the nucleus splits into two. The two parts are assisted in breaking apart by the long-range electrostatic repulsion between the protons.

Fig. 5.1
Elongated nucleus with narrow neck

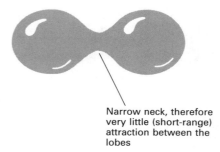

Narrow neck, therefore very little (short-range) attraction between the lobes

5.2 THE THERMAL REACTOR

When a $^{235}_{92}$U nucleus captures a neutron and undergoes fission (section 2.10) an average of about 2.5 neutrons is released. (The actual number depends on just which pair of fission products is formed.) The principle of the thermal reactor is to cause these neutrons to produce more fission by being captured by other $^{235}_{92}$U nuclei so that a **chain reaction** occurs.

In natural uranium only about 1 atom in 140 is a $^{235}_{92}$U atom – the rest are $^{238}_{92}$U. $^{238}_{92}$U can be fissioned, but only by being bombarded with very fast neutrons. On the other hand, slow neutrons are required to produce fission in $^{235}_{92}$U. The neutrons released by the fission of $^{235}_{92}$U are not fast enough to produce fission in $^{238}_{92}$U, but need to be slowed down before they can cause fission with $^{235}_{92}$U.

The neutrons are slowed down by the use of a material called a **moderator** – commonly graphite, water or heavy water (D_2O). In a graphite-moderated reactor, for example, the uranium fuel is in sealed tubes which are arranged inside a block of graphite (Fig. 5.2). The neutrons released by the fission of $^{235}_{92}$U make repeated collisions with the atoms of the moderator and are slowed to such an extent that they are far more likely to cause fission of $^{235}_{92}$U than to be unproductively captured by $^{238}_{92}$U.

Fig. 5.2
The advanced gas-cooled reactor (AGR) – an example of a graphite-moderated thermal reactor

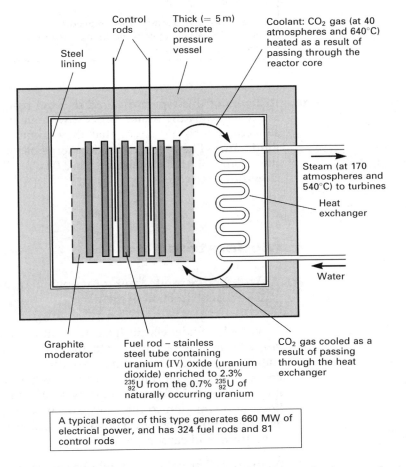

Control rods

Thick (= 5 m) concrete pressure vessel

Coolant: CO_2 gas (at 40 atmospheres and 640°C) heated as a result of passing through the reactor core

Steel lining

Steam (at 170 atmospheres and 540°C) to turbines

Heat exchanger

Water

Graphite moderator

Fuel rod – stainless steel tube containing uranium (IV) oxide (uranium dioxide) enriched to 2.3% $^{235}_{92}$U from the 0.7% $^{235}_{92}$U of naturally occurring uranium

CO_2 gas cooled as a result of passing through the heat exchanger

A typical reactor of this type generates 660 MW of electrical power, and has 324 fuel rods and 81 control rods

Ideally, exactly one neutron per fission is required to sustain the reaction; much higher rates than this would release energy too quickly and the reaction would go out of control. **Control rods** of boron-coated steel or cadmium are used to keep the net rate of production of neutrons to the required level by capturing the necessary proportion before they can initiate fission. When the control rods are moved upwards out of the heart of the reactor, the number of neutrons left to produce fission is increased; when the rods are lowered, the number of neutrons is decreased.

The heat energy produced by the fission reaction is removed by passing a coolant such as carbon dioxide or water through the reactor. The coolant then passes through some form of heat exchanger, producing steam to drive turbines which in turn generate electricity. A thick concrete shield prevents potentially harmful radiation from reaching the operators.

The energy released by the fission of a single uranium nucleus is about 200 MeV, most of which is dissipated as heat in the reactor core. Almost 90% of the energy is produced by the fission process itself; the rest is due to the radioactive decay of the various fission products (see Table 5.1). The antineutrinos which result from this decay account for 4.3% of the energy released. This is lost when the antineutrinos, which have virtually no interaction with matter, escape from the reactor.

Table 5.1
The energy released by
the fission of $^{235}_{92}$U

Fission process		
KE of fission fragments	83.5%	
KE of prompt neutrons	2.4%	89.4%
Energy of prompt γ-rays	3.5%	
Radioactive decay of fission products		
Energy of β-particles, γ-rays and delayed neutrons	6.3%	10.6%
Energy of antineutrinos	4.3%	

Note   Reactors of this type are called <u>thermal</u> reactors because the neutrons (which are released by the fission process with an average energy of about 2 MeV) are slowed to such an extent that they come into thermal equilibrium with the moderator. The average kinetic energy of the neutrons is then equal to that of the atoms of the moderator (≈ 0.04 eV). The neutrons are referred to as **thermal neutrons**.

Moderators

An effective moderator must slow neutrons down without absorbing more than a small fraction of them.

If a particle collides elastically with one much heavier than itself, it bounces off with virtually no loss of energy. On the other hand, if it collides elastically with a stationary particle of the <u>same mass</u> as itself, it is brought to rest – the whole of its energy having been transferred to the other particle. It follows that an effective moderator should contain nuclei with about the same mass as the neutron. The best approximation to this is hydrogen. Unfortunately, since hydrogen itself is gaseous, it is not a practical possibility, and instead, it is used in the form of water. Deuterium and carbon are also used.

As many as fifty collisions might be necessary to bring the neutrons into thermal equilibrium with the moderator. It is important therefore that the probability of a neutron being absorbed by the moderator in any single collision is very small.

Hydrogen in the form of water. This has a considerable disadvantage in that hydrogen nuclei (protons) have a high probability of absorbing neutrons to form deuterium nuclei (deuterons) – see Table 5.2. This is offset by using **enriched uranium**, i.e. uranium in which the normal proportion of $^{235}_{92}$U has been increased from 0.7% to about 3%.

Deuterium in the form of heavy water (D_2O). This is very effective. The deuterium nucleus produces considerable slowing and neither it nor the oxygen nucleus has a high absorption cross-section for thermal neutrons. (Table 5.2).

Carbon in the form of graphite. This is heavier than deuterium and therefore less effective in slowing the neutrons. However, it is cheap and has an acceptable absorption cross-section.

Table 5.2
Absorption cross-sections for thermal neutrons

Nucleus	Absorption cross-section (in barns)
Oxygen	2.7×10^{-4}
Deuterium	5.3×10^{-4}
Carbon	3.4×10^{-3}
Hydrogen	3.3×10^{-1}
Uranium	7.6
Boron	7.6×10^2
Cadmium	2.4×10^3

Note The greater the absorption cross-section, the greater the likelihood of absorption. Thus hydrogen nuclei are about a hundred times more likely to absorb a neutron than are carbon nuclei.

Note **The absorption cross-section** is measured in barns and is the <u>effective</u> area that the nucleus concerned presents to an incident neutron (Fig. 5.3).

Fig. 5.3
The absorption cross-section of a nucleus

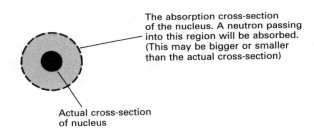

The absorption cross-section of the nucleus. A neutron passing into this region will be absorbed. (This may be bigger or smaller than the actual cross-section)

Actual cross-section of nucleus

$$1 \text{ barn} = 10^{-28} \text{ m}^2$$

Thus a nucleus which has a cross-section of 1 barn will absorb a neutron coming within an area of 10^{-28} m^2 centred on the nucleus.

Fuel Elements

The uranium fuel is packed into narrow tubes which are arranged in a regular geometrical array and embedded within the moderator. The tubes are <u>narrow</u> so that the neutrons released by the fission process have a high probability of escaping into the moderator and being slowed down before they next encounter a fuel element. The exact form in which the uranium fuel is employed depends on the particular type of reactor. In the AGR, for example, it is in the form of hollow pellets of uranium(IV) oxide (uranium dioxide) in ceramic form enriched to 2.3% $^{235}_{92}U$. The pellets are contained in stainless steel tubes to prevent the escape of the radioactive fission products and to facilitate their removal when the fuel is exhausted. Features of some common thermal reactors are given in Table 5.3. The AGR has been developed from the Magnox reactor. It is more efficient because it operates at a higher temperature, a feature made possible because the fuel elements can withstand higher temperatures than those used in Magnox reactors.

Table 5.3
Features of some
common types of thermal
reactor

Type	Fuel	Coolant	Moderator
Magnox	Uranium metal in magnesium alloy cans	CO_2 at 400 °C and 20 atm	Graphite
Advanced gas-cooled reactor (AGR)	Enriched uranium(IV) oxide (uranium dioxide) pellets in stainless steel cans	CO_2 at 640 °C and 40 atm	Graphite
Pressurized water reactor (PWR)	Enriched uranium(IV) oxide (uranium dioxide) pellets in zirconium alloy cans	Water at 315 °C and 150 atm	Water

Control of Reaction Rate

Both boron and cadmium have high absorption cross-sections for thermal neutrons (Table 5.2), and are used in the from of rods which are moved in and out of the reactor to control the reaction rate. This form of control would not be possible were it not for a piece of good fortune. Over 99% of the neutrons present (the so-called, **prompt neutrons**, see section 2.10) are released at the instant of fission. If there were no others, the fission rate could build up to a dangerously high level in as little as 0.1 s. Movement of the control rods is a very slow process in comparison and would not provide an effective means of control. Fortunately, about 0.7% of the neutrons are the result of neutron decay of some of the fission products. For example $^{87}_{35}Br$ decays by β^--emission with a half-life of 56 s to produce $^{87}_{36}Kr$. In some instances the $^{87}_{36}Kr$ is in an excited state and decays (instantly) by emitting a neutron. Thus, some of the neutrons (known as **delayed neutrons**) are produced indirectly by the relatively slow process of β^--decay. Although only 0.7% of the neutrons are produced in this way, reactors are designed so that it is these that hold the balance between a reaction which is dying away and one which is going dangerously out of control; this gives ample time for the control rods to be repositioned.

Coolant

Water, heavy water and carbon dioxide are commonly used as coolants, (see Table 5.3). The coolant absorbs neutrons and becomes radioactive as it passes through the reactor core. It is important, therefore, that the coolant is contained within a closed system.

5.3 CRITICAL MASS

If a nuclear reaction is to produce a self-sustaining chain reaction, each fission must give rise to at least one other. This is less likely to happen with a small lump of material than a large one, because a greater proportion of the neutrons produced in a small lump will be lost through the surface. (This is because the volume of a sphere, for example, increases as a function of r^3 whereas the surface area increases only as a function of r^2.)

It follows that nuclear fission results in a self-sustaining chain reaction only if the volume, and therefore the mass, of fissile material present is greater than some minimum value. This is called the critical mass, i.e.

> **The critical mass** of a fissile material is the minimum mass of it that will support a self-sustaining chain reaction.

The critical mass depends on shape – for $_{92}^{235}U$ in the form of a sphere it is of the order of 10 kg.

A reaction in which one fission causes exactly one other is said to be **critical**. If it produces more than one, it is **supercritical**; if it produces less than one, it is **subcritical**.

5.4 REACTOR SAFETY CONSIDERATIONS

Nuclear reactors are designed with a number of safety features in mind.

(i) Large quantities of potentially harmful neutrons, γ-rays and β-particles are produced. (The antineutrinos produced by the β-decay of the fission products cannot be prevented from escaping but they have so little interaction with matter that they are not harmful.) The β-particles have a limited range and are easily stopped. Neutrons and γ-rays produce very little ionization and are therefore extremely penetrating. As a consequence, extremely thick ($\approx 5\,m$) concrete shields are used to protect personnel working in the immediate vicinity of a reactor.

(ii) The coolant is at a high pressure and therefore the core is surrounded by a **pressure vessel**. In many reactors the thick concrete shield also acts as the pressure vessel.

(iii) The fission process can be stopped simply by inserting the control rods to the maximum extent. (In an emergency it is essential that this is done as quickly as possible.) This reduces the energy being generated by the reactor but it does not stop it altogether. The fission products continue to produce energy through radioactive decay – and this cannot be stopped! If the cooling system were to fail, the heat produced by this process would melt the core (**melt-down**). The reactor is designed to contain the core even in this eventuality.

(iv) Reactors are designed to 'fail-safe' in the event of an emergency such as electrical failure.

5.5 DISPOSAL OF RADIOACTIVE WASTE

Radioactive waste is classified as low-level, intermediate level or high-level according to the intensity of the radiation it produces.

Low-Level Waste

This is material which is only slightly radioactive. Solid forms, such as protective clothing and laboratory instruments which have been in contact with radioactive materials are compacted and stored in steel containers inside large concrete vaults. Liquid waste, such as cooling-pond water (section 5.6), is treated to reduce the radioactivity to a negligible level and is then discharged into the sea.

Intermediate-Level Waste

This is far less active than high-level waste. It includes discarded fuel cans, reactor components and sludges produced by various treatment processes. It is encapsulated with cement in steel drums. These are then held in concrete storage areas or are stored in underground repositories.

High-Level Waste

This comprises either spent reactor fuel or that component of it that cannot be re-used after reprocessing. The way this is dealt with is described in section 5.6.

5.6 TREATMENT OF USED REACTOR FUEL

After a few years in a nuclear reactor the fuel is depleted to such an extent that it can no longer support a chain reaction and so has to be replaced.

(i)   The spent fuel elements (fuel + can) are highly radioactive and therefore remote handling methods have to be used to remove them from the core.

(ii)   They are then stored under water in 'cooling ponds'. The water absorbs the heat generated by the radioactive decay of the fission products and protects plant personnel from the radiation that is emitted. Many of the decay products are short-lived and their activity is considerably reduced by the time the fuel elements are removed from the cooling ponds to be sent for reprocessing.

(iii)   The fuel is removed form the cans using remote controlled equipment and is then dissolved in nitric acid. It is subsequently separated into three portions:

Uranium ($^{235}_{92}$U and $^{238}_{92}$U) (96%),

Plutonium (1%),

Fission products (3%).*

The uranium is either used in the 'blankets' of fast reactors (see section 5.7) or, more often, enriched for re-use in a thermal reactor. The plutonium can be used in both thermal and fast reactors. Though it is only mildly radioactive (an α-emitter with a half-life of 2.4×10^4 years), plutonium is extremely toxic and must be treated with great care. Until recently the highly active fission products have been stored in liquid form in double-walled stainless steel tanks. Nowadays they are converted into a type of glass and stored in stainless steel flasks in air-conditioned repositories.

Note   Plutonium is produced as a result of $^{238}_{92}$U absorbing neutrons – see section 5.7.

*The percentages refer to the fuel from Magnox reactors.

5.7 THE FAST BREEDER REACTOR

When plutonium 239 ($^{239}_{94}$Pu) captures a neutron to become $^{240}_{94}$Pu the $^{240}_{94}$Pu is in a highly excited state and undergoes fission releasing an average of 2.9 neutrons in the process. (The actual number depends on just which pair of fission fragments is produced.)

Some of the neutrons are captured by other $^{239}_{94}$Pu nuclei and initiate further fissions so that a chain reaction results. **There is no moderator** – the plutonium is fissioned by fast neutrons – hence 'fast breeder reactor'.

Plutonium does not occur naturally. It is produced when $^{238}_{92}$U absorbs a neutron to become $^{239}_{92}$U. The sequence of events is

$$^{238}_{92}\text{U} + ^{1}_{0}\text{n} \rightarrow ^{239}_{92}\text{U} \rightarrow ^{239}_{93}\text{Np} + ^{0}_{-1}\text{e} + \bar{v}$$

$$^{239}_{93}\text{Np} \rightarrow ^{239}_{94}\text{Pu} + ^{0}_{-1}\text{e} + \bar{v}$$

(Neither $^{239}_{92}$U nor $^{239}_{93}$Np occur in nature. They have half-lives of 23.5 minutes and 2.35 days respectively. $^{239}_{94}$Pu is much more stable – it has a half-life of 2.4×10^4 years.)

The central core of the reactor, containing the plutonium fuel, is surrounded by a blanket of natural uranium, of which 99.3% is $^{238}_{92}$U. Some of the neutrons which escape from the core are absorbed by $^{238}_{92}$U nuclei to create more $^{239}_{94}$Pu. Since reactors of this type can produce more plutonium than they use, they are known as breeder reactors. Because they rely on $^{238}_{92}$U, rather than the much rarer $^{235}_{92}$U, they make much more efficient use of the world's supply of uranium than do thermal reactors.

Fast breeder reactors present greater design problems than thermal reactors and they have not yet advanced beyond the prototype stage. **Liquid sodium** is used as the coolant because the core has to be compact and therefore the heat generated in it needs to be removed very efficiently.

5.8 FUSION REACTORS

The reader should be familiar with section 2.11 before continuing.

The Deuterium–Tritium (D–T) Reaction

The reaction most likely to be used in the first working fusion reactors involves fusing deuterium ($^{2}_{1}$H) and tritium ($^{3}_{1}$H) to produce helium 4.

$$^{2}_{1}\text{H} + ^{3}_{1}\text{H} \rightarrow ^{4}_{2}\text{He} + ^{1}_{0}\text{n} + 17.6\,\text{MeV}$$

The world has an abundant supply of deuterium; it constitutes 0.015% of naturally occurring hydrogen and can easily be extracted from water. Tritium is radioactive and does not occur naturally, but it could be produced by surrounding the core of the reactor with a blanket of lithium (which is plentiful) and allowing this to absorb neutrons from the D–T reaction (Fig. 5.4). Lithium occurs as lithium 6 (7.5%) and lithium 7 (92.5%); the **tritium-producing reactions** are:

$$^{6}_{3}\text{Li} + ^{1}_{0}\text{n} \rightarrow ^{3}_{1}\text{H} + ^{4}_{2}\text{He}$$

and

$$^{7}_{3}\text{Li} + ^{1}_{0}\text{n} \rightarrow ^{3}_{1}\text{H} + ^{4}_{2}\text{He} + ^{1}_{0}\text{n}$$

Fig. 5.4
Schematic diagram of a
fusion reactor

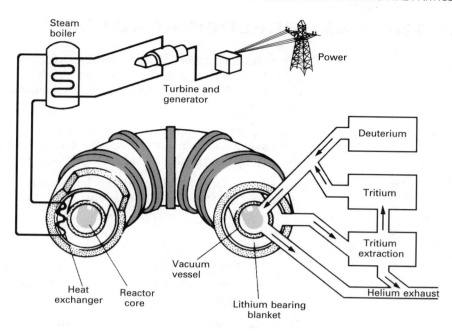

The tritium would be removed from the blanket for later use, i.e. **the reactor would breed its own tritium**.

The kinetic energy of the neutrons accounts for 80% of the energy released by the D–T reaction. The neutrons would be absorbed by the lithium, heating it in the process. Some form of heat exchanger could be used to extract this heat to produce steam to drive turbines and generate electricity.

Plasma Confinement

The practical realization of generating power by controlled thermonuclear fusion requires that some means is found of confining the plasma: **(i) long enough, (ii) at a sufficiently high temperature and (iii) at a sufficiently high density**, for a self-sustaining fusion reaction to occur.

The plasma cannot be allowed to touch the walls of any vessel that might be used to contain it – if it did, it would rapidly lose temperature and would also release impurity atoms from the walls which would contaminate the plasma and produce further cooling. (Note, contrary to what is often supposed, the plasma would not melt the containing vessel – its heat capacity would be far too small.)

Gravitational Confinement

In the Sun (and the other stars) the plasma is held together by gravitational attraction. This is possible only with plasmas of comparable mass, and cannot be used on Earth. However, there are two possibilities which are currently under investigation – **magnetic confinement** and **inertial confinement**. In magnetic confinement, the plasma density is relatively low ($\sim 10^{-3}\,\mathrm{kg\,m^{-3}}$) but the confinement time is long ($\sim 1\,\mathrm{s}$). In inertial confinement, the plasma density is very much higher but for a very much shorter time ($\sim 10^{-9}\,\mathrm{s}$).

Magnetic Confinement

The systems of this type that currently show the most promise are based on the Soviet **Tokamak** principle. The plasma is contained within a **torus** (i.e. a hollow tube in the form of a ring – like a hollow doughnut) in which a helical magnetic field is generated (Fig. 5.5). The plasma particles (electrons and ions) orbit the torus by spiralling around the field lines and are therefore held clear of the walls.

Fig. 5.5
Plasma confinement
using a torus

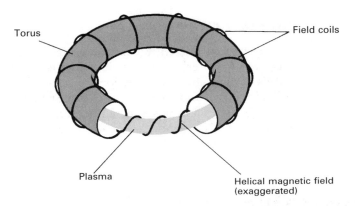

Note. The helical magnetic field is due to the combined effect of the field produced by the field coils and that created by the plasma current

Torus

Field coils

Plasma

Helical magnetic field
(exaggerated)

The world's largest and most advanced Tokamak project is the **Joint European Torus (JET)** at Culham near Oxford. The knowledge gained from this experiment should eventually lead to a working reactor. It is envisaged that the plasma in a reactor will be created by inducing a current of several million amperes in a deuterium–tritium gas mixture contained within the torus. This same current will then heat the plasma. As the plasma heats up its resistance will fall, and this particular form of heating will be ineffective by the time the temperature has

The interior of the Joint European Torus (JET).

reached 40×10^6–50×10^6 K. This is high enough for some D–T fusion reactions to occur, but not at a sufficiently high rate for there to be a net gain of energy. It will therefore be necessary to provide additional forms of heating such as **radio-frequency heating**, in which the plasma simply absorbs the energy of high-power radio waves. The highly energetic helium nuclei produced by the fusion reaction will be trapped inside the plasma by the magnetic field and will contribute even more heating.

As the fuel is used up, it will need to be replenished. This could be achieved by injecting pellets of frozen deuterium and tritium at high speed into the heart of the plasma.

Inertial Confinement

In one system currently being investigated a tiny (1 mm) pellet containing a mixture of deuterium and tritium fuel is struck simultaneously by short (1 ns) pulses of high-intensity radiation from several infrared lasers. The surface of the pellet vaporizes and creates a shock wave which increases the density of the core by up to 10^3 times and raises its temperature to over 10^8 K. This occurs in less than 1.5 ns, which is too short a time for the particles to disperse, i.e. they are confined by their own inertia!

5.9 ADVANTAGES OF FUSION

(i)   **The fuels required are plentiful,** and furthermore (unlike oil, for example) they are widely available throughout the world. There is a virtually unlimited supply of deuterium (which can be extracted from water) and it is estimated that the known reserves of lithium (which is required for the production of tritium) will last for hundreds of years, by which time fusion reactors using the D–D reaction should have been developed.

(ii) **A fusion reactor is inherently safe**. It is impossible for the reaction to go out of control and cause an explosion because there is very little fuel present in the reactor at any one time. The reaction is self-sustaining only under optimum conditions and the reaction therefore automatically 'fails-safe' – if any form of malfunction occurs, the plasma simply collapses.

(iii) **There are no long-lived radioactive waste products**. The only radioactive product of the fusion reaction is tritium, and this is re-used. The structure of the reactor becomes radioactive through absorbing neutrons, but the problems associated with this can be reduced by careful selection of the various materials used.

(iv) **There is no atmospheric pollution**.

CONSOLIDATION

Liquid-Drop Model of Fission

When a $^{235}_{92}$U nucleus absorbs a neutron it oscillates and a narrow neck forms between two approximately spherical lobes. The long-range electrostatic repulsion between the two lobes exceeds the short-range attractive forces and the nucleus splits into two parts.

Thermal Reactor

$^{235}_{92}$U is fissioned by **thermal neutrons**, i.e. neutrons which have been slowed to such an extent that they are in thermal equilibrium with the reactor core.

Moderator – <u>slows</u> the neutrons so that they are more likely to cause fission.

Control rods – <u>absorb</u> neutrons so that the reaction does not go out of control.

Coolant – carries away the heat generated in the reactor core.

Fast Breeder Reactor

$^{239}_{94}$Pu is fissioned by <u>fast</u> neutrons.

There is no moderator.

$^{238}_{92}$U absorbs neutrons produced by the fission process to create (breed) more $^{239}_{94}$Pu.

The critical mass of a fissile material is the minimum mass of it that will support a self-sustaining chain reaction.

Thermonuclear Fusion

To date no one has succeeded in producing a controlled **thermonuclear fusion reaction** which has been self-sustaining. Methods currently being investigated involve **magnetic confinement** and **inertial confinement**.

Thermonuclear fusion occurs naturally in the Sun and other stars.

QUESTIONS ON CHAPTER 5

1. (a) Explain what is meant by a chain reaction in nuclear fission. For the reaction to proceed, why must the mass of fissile material exceed a certain critical value and why does this mass depend on the shape of the sample?

(b) In a thermal neutron fission reactor explain the function of
   (i) the moderator,
   (ii) the control rods,
   (iii) the coolant.
   In each case suggest *one* suitable material, and state *two* of its essential physical properties. [J]

2. (a) Discuss in detail the processes by which energy is released in the fission of the $^{235}_{92}$U content of the fuel rods in a graphite moderated reactor, and explain how it is dissipated.

(b) When a neutron collides head on and elastically with a stationary proton, nearly all the energy of the neutron is transferred to the proton. Discuss the application of this result to the choice of a material as moderator in a fission reactor. Indicate the features of the interactions of neutrons with carbon nuclei which lead to the use of graphite as moderator in power generating reactors. [J]

3. (a) Explain the action of a moderator in a thermal neutron fission reactor. Give *two* physical factors which are important in making a choice of moderator material.

(b) Discuss why cadmium is commonly used to make control rods in such a reactor. Explain how control rods are used.

(c) Spent fuel elements from a nuclear reactor contain appreciable concentrations of many radioactive isotopes. Explain why this is so. [J]

4. (a) When ^{235}U is bombarded by neutrons, ^{236}U can be formed, undergo fission and produce ^{140}Xe and ^{93}Sr. When one atom undergoes fission in this way, 195 MeV of energy is released. If the nuclear binding energy of both ^{140}Xe and ^{93}Sr is 8.5 MeV per nucleon, calculate the binding energy per nucleon of ^{236}U.

(b) When such a fission takes place in a fuel rod in a nuclear reactor, about 175 MeV of this energy is released immediately, followed by emission of the remaining energy over a period of time. List the means by which both prompt and delayed energy is carried away, and discuss whether it is all eventually converted into useful heat.

(c) When a neutron collides head on with a carbon atom, it can be shown that

$$\frac{\text{Residual energy of neutron}}{\text{Original energy of neutron}} = 0.7$$

Show that about 49 of these collisions are necessary to slow down a fission neutron of energy 1 MeV to thermal energy, approximately 0.03 eV.

State why the moderation process is necessary in a nuclear reactor. Since so many collisions are involved, what other property of a moderator is essential, apart from the ability to slow down neutrons?

[J]

5. (a) Discuss how the nuclear chain reaction in a power reactor differs when the reactor is
(i) in normal operation and
(ii) starting to run out of control.
Describe the part played by the neutrons in each case.

(b) Explain the following facts concerning fission reactors, including a description of the type of radioactivity to be expected and why it arises.
(i) The fuel rods are only slightly radioactive when they are manufactured, but they become highly radioactive after they have been in use.
(ii) Shielding material around the core is stable when the reactor is constructed, but becomes radioactive as the reactor is used.

(c) Describe the stages of treatment of the spent fuel rods after they have been removed from a reactor, giving details of how the active wastes are dealt with.

[J, '93]

6. (a) Write notes on the safety aspects of the use of nuclear reactors for power production under the following headings.
(i) Normal operation.
(ii) Emergency shut down.
(iii) Waste disposal, including the treatment of spent fuel rods.

(b) There is evidence that samples of uranium mined at Gabon in Africa have acted as fuel in a natural nuclear reactor. State *one* difference you would expect between purified samples of this uranium and purified samples of uranium mined elsewhere. Explain your reasoning. [J]

7. (a) Explain what is meant in nuclear fission by a *self-sustaining chain reaction*.

(b) Show that, when a spherical mass of fissile material of radius R is undergoing fission,

$$\frac{\text{Rate of loss of neutrons}}{\text{Rate of production of neutrons}} \propto \frac{1}{R}.$$

Assume that the only neutrons absorbed by the material are those that cause fission.
Explain why this result places a limit on the size of the sample of material required to establish a self-sustaining chain reaction in fissile material.

(c) Explain why it is possible for newly-canned fuel rods to be handled by personnel, whilst elaborate safety precautions and remote-control devices are needed when spent fuel rods are removed from a nuclear reactor.

[J]

8. (a) The processes of *moderation, control,* and *cooling* are essential to the operation of a thermal reactor. For the terms printed in italics
(i) give a brief description of *each* process and explain how *each* process is achieved,
(ii) name a suitable material used to achieve *each* process and indicate why it is suitable.

(b) State *two* principal design differences between the first generation (Magnox) British nuclear reactors and the second generation (AGR) advanced gas-cooled reactors.

(c) When a uranium nucleus undergoes fission, about 200 MeV of energy is released. For a nuclear power station which has an electrical power output of 1200 MW, and which converts 40% of the fission energy into electrical energy, find
(i) the number of uranium nuclei which undergo fission per second,
(ii) the mass of fuel which is changed into energy during 24 hours of continuous operation. ($e = 1.6 \times 10^{-19}$ C, $c = 3.0 \times 10^8 \text{ m s}^{-1}$.) [J, '91]

9. When ^{235}U undergoes fission in a nuclear reactor, the average number of neutrons produced per fission is about 2.5.

(a) Describe what happens to these neutrons in a graphite-moderated reactor which is being used to produce power. Describe briefly the arrangement of the fuel in the core of such a reactor, stating how the heat is removed from the core.

(b) Explain how the reaction is controlled, stating two essential properties of the control rod material.   [J]

10. (a) Explain how a chain reaction can be maintained in a fission reactor using natural uranium as fuel, and how it is controlled so that energy is produced at a constant rate.

(b) It is possible to obtain energy from the fusion of light nuclei, but so far it has proved impossible to use nuclear fusion for the large-scale generation of electrical energy.

(i) Give an equation for one suitable fusion reaction.

(ii) State *two* of the main difficulties involved in producing controlled fusion.

(iii) Indicate briefly *one* method of producing fusion reactions which has been studied on a laboratory scale.   [J]

11. (a) Explain those features of each of the following which ensure the safety of personnel and the public when a nuclear reactor is used to supply power:

(i) the shielding,

(ii) the pressure vessel,

(iii) the control rods.

State what materials are used in each case in a graphite moderated fission reactor.

(b) Indicate the chief stages in the treatment of spent fuel rods from such a reactor, giving reasons for these stages.

(c) Give *two* potential advantages concerned with safety which the production of nuclear energy by a fusion process might have over corresponding fission processes.   [J]

12. It is possible in principle to obtain energy from the fusion of light nuclei. One suitable reaction is

$$^3H + {}^2H \rightarrow X + Y + Q$$

where X and Y represent the two products of the reaction and Q the energy release. Where necessary using data from the table given below, answer the following questions.

(a) Identify the products X and Y.

(b) Calculate the value of Q in MeV.

(c) Calculate the kinetic energy of each of the product nuclei, X and Y, assuming that the initial nuclei are at rest.

(d) Give two physical reasons why it has so far proved impossible to build a practical power station using fusion as an energy source.

In the following table, which lists a selection of nuclides, the mass of an isotope is given for a neutral atom of the substance and is quoted in unified atomic mass units, u. 1 u is equivalent to 931.3 MeV.

Name	Symbol	Charge number, Z	Mass number, A	Atomic mass, M
Electron	$e^{\pm}$	± 1	—	0.00055
Proton	p	1	1	1.00728
Neutron	n	0	1	1.00867
Hydrogen	^{1}H	1	1	1.00783
	^{2}H	1	2	2.01410
	^{3}H	1	3	3.01605
Helium	^{3}He	2	3	3.01603
	^{4}He	2	4	4.00260
Lithium	^{6}Li	3	6	6.01513
	^{7}Li	3	7	7.01601

[J]

13. (a) Explain what is meant by a chain reaction which uses nuclear fission. For the reaction to proceed why must the mass of fissile material exceed a certain critical value, and why does this critical mass depend on the ratio: *surface area/volume?*

For a particular fissile material, what shape will give the smallest critical mass?

(b) One possible method of producing energy by fusion involves several reactions which can be summarised by the following equation.

$$5\,^2H \rightarrow {}^1H + {}^3He + 2n + {}^4He + Q$$

Q represents the energy released.

(i) Using data selected from the table in Question 12 calculate the energy liberated, in MeV, when 1.0 kg of deuterium (^{2}H) is converted.

(ii) Calculate the mass of ^{235}U required to produce the same amount of energy, if the fission of one atom of ^{235}U produces, on average, 200 MeV. (Avogadro constant $= 6.0 \times 10^{23}$ mol^{-1}.)

(c) Give *three* advantages which such a fusion process would have as a source of power compared with a fission process. [J]

14. State where nuclear fusion occurs continuously in nature and explain how the physical effect that inhibits fusion is overcome in this instance. How may fusion be achieved experimentally under conditions different from those applying to natural fusion? [J]

15. (a) When a nuclear power station is commissioned, useful energy is first produced when the reactor becomes *critical*.

    (i) Identify the principal components of a power reactor (as used in an electrical generating station) and state their functions.

    (ii) Explain what happens when a reactor passes from a sub-critical to the critical state.

(b) Explain why spent fuel rods from a reactor are highly radioactive.

(c) Outline the precautions that are taken in dealing with the spent fuel rods

    (i) in the short term,

    (ii) in the long term.

(d) The following equation represents a fusion reaction between two nuclei of deuterium ($^2_1$H):

$$^2_1\text{H} + {}^2_1\text{H} \rightarrow {}^3_2\text{He} + {}^1_0\text{n} + \text{Energy}\,(E)$$

The masses of the particles involved in this reaction (expressed in atomic mass units, u) are

$^2_1$H: 2.01410 u $^3_2$He: 3.01603 u

$^1_0$n: 1.00867 u

Calculate the energy (E) produced in
(i) joule (J),
(ii) mega-electronvolt (MeV).
($1\,\text{u} = 1.66 \times 10^{-27}$ kg, $c = 3.00 \times 10^8\,\text{m s}^{-1}$, $e = 1.60 \times 10^{-19}$ C.) [O&C, '91]

16. (a) The diagram is a cross-section through the core of a nuclear reactor. Parts P are fixed in position, parts Q can be raised and lowered during the operation of the reactor and parts R are removed and replaced from time to time during the lifetime of the reactor.

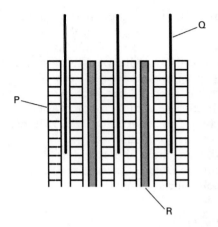

What are P, Q and R? Explain the function of each when the reactor is releasing nuclear energy at a controlled rate.

(b) One of the possible nuclear reactions which occur in the fission of uranium 235 is

$$^{235}_{92}\text{U} + {}^1_0\text{n} \rightarrow {}^{138}_{55}\text{Cs} + {}^{96}_{37}\text{Rb} + 2{}^1_0\text{n}.$$

Use the following information to show that the energy released by this reaction is 2.7×10^{-11} J. Hence calculate the energy available from the fission of 1.00 kg of uranium 235.

(Avogadro constant $= 6.03 \times 10^{23}$ mol^{-1}, speed of light in vacuo $= 3.00 \times 10^8\,\text{m s}^{-1}$, unified atomic mass unit (u) $= 1.66 \times 10^{-27}$ kg, mass of neutron $= 1.009$u, masses of relevant nuclei: U-235 $= 235.044$u, Cs-138 $= 137.920$u, Rb-96 $= 95.932$u.) [L, '91]

17. (a) Describe what is meant by a thermonuclear fusion process.
Explain why such a process is difficult to:
(i) initiate,
(ii) control
under laboratory conditions.

(b) One such process in the Sun involves the fusion of four hydrogen nuclei (each of mass 1.0078 u) to form one helium nucleus (of mass 4.0026 u). The masses of other reaction products are negligible.

$$4{}^1_1\text{H} \rightarrow {}^4_2\text{He} + 2{}^{\ \ 0}_{-1}\text{e}$$

Given that $1\,\text{u} = 1.67 \times 10^{-27}$ kg and the speed of light $c = 3.00 \times 10^8\,\text{m s}^{-1}$, calculate how much energy (in joule) is released when a single helium nucleus is produced. [O, '93]

6

PARTICLE PHYSICS

6.1 INTRODUCTION

Particle physics is concerned with particles such as protons, neutrons, electrons, neutrinos and the various mesons, i.e. particles which are about the size of a nucleon or smaller. They are often referred to as '**fundamental particles**' – some of them genuinely are fundamental but some, as we shall see, are composed of other particles and are therefore not fundamental in the strict sense.

6.2 PARTICLES AND ANTIPARTICLES

Each type of particle has an antiparticle. **Antiparticles are not constituents of ordinary matter**. They can be created when particles collide in high-energy accelerators, or when cosmic rays interact with matter, or as a result of radioactive decay.

(i)   A particle has the same mass as its antiparticle.

(ii)   A particle and its antiparticle have equal and opposite charge.

(iii)   An <u>unstable</u> particle and its antiparticle have the same lifetime.

Properties of three particles together with those of their antiparticles are given in Table 6.1.

Table 6.1
The electron, proton and neutrino and their anti-particles

Particle Antiparticle	Symbol	Charge	Rest mass (MeV/c^2)*	Lifetime
Electron	e^-	-1	0.511	Stable
Positron	e^+	1		
Proton	p	1	938.3	Stable
Antiproton	$\bar{p}$	-1		
Neutrino	v	0	Probably zero	Stable
Antineutrino	$\bar{v}$	0		

*The use of MeV/c^2 as a unit of mass is discussed in section 6.3.

Notes   (i)   These particular antiparticles are <u>stable</u> in that they would exist for ever <u>in isolation</u>. In practice they are likely to encounter their particle counterparts and become annihilated a short time after being created (see section 6.4).

(ii)   There are actually three types of neutrino (see section 6.8).

(iii)   Some <u>neutral</u> particles and their antiparticles are identical (e.g. the photon and the π^0 meson); others are not – just how they differ will become apparent in the sections that follow.

6.3 RELATIVISTIC EQUATIONS IN PARTICLE PHYSICS

According to the special theory of relativity, the mass of a body is not constant but increases with velocity. A body moving with velocity v has a mass m, its **relativistic mass**, given by

$$m = \frac{m_0}{\sqrt{1 - \dfrac{v^2}{c^2}}}$$

[6.1]

where m_0 is its **rest mass** (i.e. its mass when it is stationary); c is the speed of light.

The increase in mass (Fig. 6.1) is not noticeable at 'ordinary' speeds – even at 10% of the speed of light the increase in mass is only 0.5%. However, the effect becomes more and more marked as the speed of light is approached. For example, a particle whose rest mass is m_0 would have a (relativistic) mass of $7.1 m_0$ at $0.99c$ and $71 m_0$ at $0.9999c$. Speeds such as these are common in particle physics.

Fig. 6.1
Relativistic mass as a
function of speed

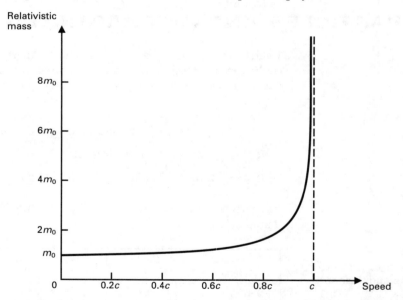

Another consequence of the theory of relativity is that mass and energy are equivalent – a mass m is equivalent to an amount of energy E where

$$E = mc^2$$

[6.2]

The energy of a body which is at rest is simply the energy associated with its rest mass and is called its **rest energy**, E_0. By equation [6.2]

$$E_0 = m_0 c^2$$

The total energy, E, of a <u>moving</u> body is the sum of its rest energy and its relativistic kinetic energy*. It is the energy associated with its relativistic mass and is given by

*When $v \ll c$, relativistic kinetic energy $\approx \frac{1}{2} mv^2$.

$$E = mc^2 = \frac{m_0 c^2}{\sqrt{1 - \dfrac{v^2}{c^2}}}$$ [6.3]

The factor $1/\sqrt{1 - (v^2/c^2)}$ is called the **Lorentz factor** and is often denoted by γ, in which case $E = \gamma m_0 c^2$ etc.

The momentum, p, of the body is given by

$$p = mv = \frac{m_0 v}{\sqrt{1 - \dfrac{v^2}{c^2}}}$$ [6.4]

By equations [6.3] and [6.4]

$$p = \frac{Ev}{c^2}$$ [6.5]

Though it is not apparent, it also follows from equations [6.3] and [6.4] that

$$E^2 = m_0{}^2 c^4 + p^2 c^2$$ [6.6]

Particles of Zero Rest Mass

It follows from equation [6.6] that

$$E = pc \qquad \left(\begin{array}{l}\text{For a particle of} \\ \text{zero rest mass}\end{array}\right)$$

It then follows from equation [6.5] that

$$v = c \qquad \left(\begin{array}{l}\text{For a particle of} \\ \text{zero rest mass}\end{array}\right)$$

Units

In particle physics it is often convenient to express energy in GeV⋆ (or MeV), momentum in GeV/c (or MeV/c) and mass in GeV/c^2 (or MeV/c^2). This forms an entirely consistent set of units and simplifies the use of equations [6.5] and [6.6]. Consider the examples that follow.

EXAMPLE 6.1

Find (a) the total energy, (b) the speed, of a particle whose rest mass is 3.0 GeV/c^2 and whose momentum is 4.0 GeV/c.

Solution

(a) $E^2 = m_0{}^2 c^4 + p^2 c^2$ (equation [6.6])

Here

$$m_0 = 3.0\,\text{GeV}/c^2 \quad \therefore \quad m_0 c^2 = 3.0\,\text{GeV}$$

and

$$p = 4.0\,\text{GeV}/c \quad \therefore \quad pc = 4.0\,\text{GeV}$$

⋆1 GeV $= 1 \times 10^9$ eV. The electronvolt (eV) is discussed in section 2.1.

By equation [6.6]

$$E^2 = 3.0^2 + 4.0^2$$

$$\therefore \quad E = 5.0 \, \text{GeV}$$

(b) $\qquad p = \dfrac{Ev}{c^2} \qquad$ (equation [6.5])

$$\therefore \quad pc = \dfrac{Ev}{c}$$

$$\therefore \quad 4.0 = \dfrac{5.0v}{c}$$

i.e. $\quad v = 0.8c$

EXAMPLE 6.2

A K^0 meson, which is at rest, decays to produce two π^0 mesons. Find (a) the total energy of each π^0, (b) the momentum of each π^0.
(Rest mass of $K^0 = 498 \, \text{MeV}/c^2$, of $\pi^0 = 135 \, \text{MeV}/c^2$.)

Solution

(a) The K^0 meson is at rest and therefore its total energy is equal to its rest energy, 498 MeV. In order that momentum is conserved, the two π^0 mesons must move in opposite directions with the same momentum as each other. Since they also have the same rest mass as each other, it follows that the total energy of the K^0 meson is shared equally between them, i.e. total energy of each $\pi^0 = \frac{1}{2}(498) = 249 \, \text{MeV}$.

(b) If the momentum of each π^0 is p, then by

$$E^2 = m_0^2 c^4 + p^2 c^2$$

$$249^2 = 135^2 + p^2 c^2$$

$$\therefore \quad pc = 209 \, \text{Mev}$$

i.e. Momentum of each $\pi^0 = 209 \, \text{MeV}/c$

EXAMPLE 6.3

An hypothetical particle, A, is at rest and decays to produce two other (hypothetical) particles, B and C. Find the total energy of each of B and C. (Rest mass of A $= 2.0 \, \text{GeV}/c^2$, of B $= 1.3 \, \text{GeV}/c^2$ and of C $= 0.3 \, \text{GeV}/c^2$.)

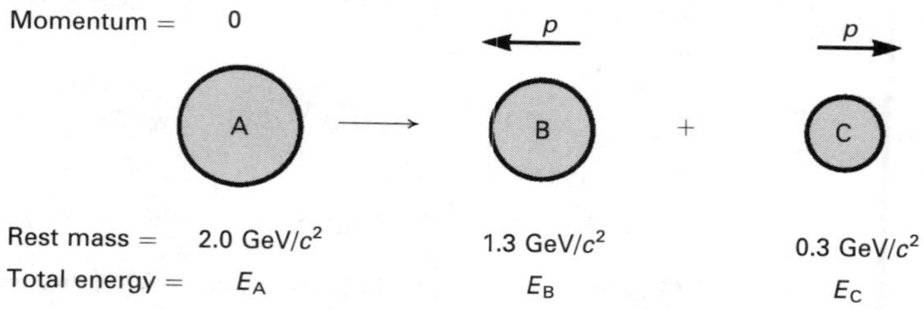

Fig. 6.2
Diagram for Example 6.3

Momentum = 0	p	p
A	B	+ C
Rest mass = 2.0 GeV/c^2	1.3 GeV/c^2	0.3 GeV/c^2
Total energy = E_A	E_B	E_C

Solution

Refer to Fig. 6.2. If the total energy of A is denoted by E_A etc., conservation of mass–energy gives

$$E_A = E_B + E_C$$

$$\therefore \quad 2.0 = E_B + E_C \qquad [6.7]$$

Since A is at rest, B and C have equal and opposite momenta; if its magnitude is p, then by equation [6.6]

$$E_B^2 = 1.3^2 + p^2c^2 \quad \text{and} \quad E_C^2 = 0.3^2 + p^2c^2$$

Eliminating p^2c^2 gives

$$E_B^2 - E_C^2 = 1.3^2 - 0.3^2 = 1.6$$

$$\therefore \quad (E_B - E_C)(E_B + E_C) = 1.6$$

Therefore by equation [6.7]

$$E_B - E_C = 0.8 \qquad [6.8]$$

Adding equations [6.7] and [6.8] gives

$$2E_B = 2.0 + 0.8 \quad \text{i.e. } E_B = 1.4\,\text{GeV}$$

Therefore by equation [6.7]

$$2.0 = 1.4 + E_C \quad \text{i.e. } E_C = 0.6\,\text{GeV}$$

QUESTIONS 6A

1. A particle whose rest mass is $5.0\,\text{MeV}/c^2$ has a momentum of $12.0\,\text{MeV}/c$. Find (a) its rest energy (in MeV), (b) its total energy (in MeV), (c) its relativistic KE (in MeV), (d) its speed (in terms of the speed of light, c). (e) On the basis of $p = mv$, find the relativistic mass of the particle in terms of MeV/c^2. What do you notice about the answers to (b) and (e)?

2. A Λ^0 particle which is at rest decays to produce a neutron and a π^0 meson. What is the total kinetic energy (in MeV) of the decay products?

(Rest mass of $\Lambda^0 = 1116\,\text{MeV}/c^2$, of $n = 940\,\text{MeV}/c^2$, of $\pi^0 = 135\,\text{MeV}/c^2$.)

3. A π^0 meson at rest decays into two γ-ray photons. What is (a) the energy (in MeV), (b) the momentum (in MeV/c) of each of these photons? (Rest mass of $\pi^0 = 135.0\,\text{MeV}/c^2$.)

4. An Ω^- particle which is at rest decays to Λ^0 and K^-. Find (a) the total energy of each of Λ^0 and K^-, (b) the kinetic energy of each of Λ^0 and K^-. (Rest mass of $\Omega^- = 1672\,\text{MeV}/c^2$, of $\Lambda^0 = 1116\,\text{MeV}/c^2$ and of $K^- = 494\,\text{MeV}/c^2$.)

6.4 PARTICLE–ANTIPARTICLE ANNIHILATION

When an electron and a positron (or a proton and an antiproton, etc.) collide they may annihilate each other.* They cease to exist and two (or occasionally three) γ-ray photons may be created in their place. (It is impossible for there to be just one photon for this would not allow <u>both</u> energy and momentum to be conserved.)

If sufficient energy is available, particles other than photons may be created. For example, an electron and a positron with a total energy in excess of $211.4\,\text{MeV}$** may annihilate to produce two muons according to

$$e^- + e^+ \rightarrow \mu^- + \mu^+$$

Note that charge is conserved. Lepton number (see section 6.8) is also conserved.

*It is also possible for a positron to be captured by an electron to form a short-lived 'atom' known as **positronium** in which the electron and the positron rotate about their common centre of mass.
**The rest mass of a muon is $105.7\,\text{MeV}/c^2$.

One of the two end caps of the OPAL detector at CERN, Geneva. OPAL is one of four giant particle detectors used on the Large Electron–Positron collider (LEP) to investigate electron–positron annihilation at particle energies of 50 GeV.

At higher energies particles of even larger rest mass can be created. The J/ψ meson, the discovery of which provided confirmation of the existence of charmed quarks (see section 6.13), was produced by electron–positron annihilation at an energy of about 3 GeV. The even more massive Υ (upsilon) meson, containing bottom quarks, results from electron–positron collisions at about 10 GeV. The carriers of the weak force, the $W^{\pm}$ and Z^0 bosons (section 6.6), were discovered as a result of proton–antiproton annihilation at energies around 90 GeV.

Note   Annihilation processes must, of course, conserve momentum. For the sake of simplicity the worked Examples and Questions that follow are restricted to situations in which the total momentum before annihilation and the total momentum after annihilation are both zero.

EXAMPLE 6.4

An electron and positron, each of which has negligible kinetic energy, annihilate and produce two γ-ray photons, each of frequency f. Calculate (a) the energy released, (b) the frequency, f. (Rest mass of electron = rest mass of positron = 9.11×10^{-31} kg, $c = 3.00 \times 10^8$ m s^{-1}, $h = 6.63 \times 10^{-34}$ J s.)

Solution

(a)   The electron and positron have negligible kinetic energy and therefore the energy released by the annihilation is equal to the rest energy of the electron plus that of the positron. Therefore

$$\text{Energy released} = 2 \times 9.11 \times 10^{-31} \, (3.00 \times 10^8)^2$$
$$= 1.64 \times 10^{-13} \, \text{J}$$

(b)   Since the energy of each photon is hf,

$$1.64 \times 10^{-13} = 2\,hf$$
$$= 2 \times 6.63 \times 10^{-34} f$$

$$\therefore \quad f = 1.24 \times 10^{20} \, \text{Hz}$$

EXAMPLE 6.5

An electron and a positron annihilate releasing 800 MeV and creating two muons (particle and antiparticle) moving in opposite directions at the same speed. Find the momentum, p, of each muon in MeV/c.

(Rest mass of each muon $=$ 105.7 MeV/c^2.)

Solution

The total energy of each muon is given by equation [6.6] as E where

$$E = (m_0{}^2 c^4 + p^2 c^2)^{1/2}$$

The energy released by the annihilation is shared equally between the muons (because they have the same speed) and therefore

$$E = 400 \text{ MeV}$$

Also

$$m_0 c^2 = 105.7 \text{ MeV}$$

Therefore

$$400 = (105.7^2 + p^2 c^2)^{1/2}$$

$\therefore$ $\quad pc = 386 \text{ MeV}$

i.e. $\quad p = 386 \text{ MeV}/c$

EXAMPLE 6.6

An electron and a positron annihilate releasing 600 MeV and creating two muons moving in opposite directions at the same speed, v. Find v in terms of c. (Rest mass of each muon $=$ 105.7 MeV/c^2.)

Solution

The total energy of each muon is given by equation [6.3] as E, where

$$E = \frac{m_0 c^2}{\sqrt{1 - \dfrac{v^2}{c^2}}}$$

Therefore

$$300 = \frac{105.7}{\sqrt{1 - \dfrac{v^2}{c^2}}}$$

$\therefore$ $\quad 1 - \dfrac{v^2}{c^2} = \left(\dfrac{105.7}{300}\right)^2$

i.e. $\quad v = 0.936c$

QUESTIONS 6B

1. An electron and a positron annihilate producing two identical γ-ray photons of frequency f. Find f if the electron and the positron are each moving (in opposite directions) at (a) $0.900c$, (b) $0.990c$. (Rest mass of electron = rest mass of positron = $0.511\,\text{MeV}/c^2$, $e = 1.6 \times 10^{-19}\,\text{C}$, $h = 6.63 \times 10^{-34}\,\text{J}\,\text{s}$.)

2. An electron and a positron are moving in opposite directions, each with a momentum of $0.400\,\text{MeV}/c$. They annihilate, producing two γ-ray photons. (a) Explain why the photons must be identical. (b) Calculate the frequency of the photons.
   (Rest mass of electron = rest mass of positron = $0.511\,\text{MeV}/c^2$, $e = 1.60 \times 10^{-19}\,\text{C}$, $h = 6.63 \times 10^{-34}\,\text{J}\,\text{s}$.)

3. A proton and an antiproton, each of which has negligible kinetic energy, annihilate creating six π^0 mesons all of which are moving at the same speed. Find the speed of the mesons, in terms of c. (Rest mass of p = rest mass of $\bar{\text{p}}$ = $938.3\,\text{MeV}/c^2$, rest mass of π^0 = $135.0\,\text{MeV}/c^2$.)

6.5 PAIR CREATION

This is the process in which a γ-ray photon ceases to exist, creating an electron–positron pair in its place. It can be shown that another particle, an atomic nucleus, for example, must also be present in order that <u>both</u> energy and momentum are conserved. The γ-ray interacts with this nucleus, causing it to recoil with the required amount of momentum. Because the mass of the nucleus is very much greater than the combined mass of the electron and positron, the nucleus takes very little of the energy. To a good approximation, therefore, the minimum photon energy required for pair creation (when a nucleus is involved) is equal to the rest energy of the electron–positron pair, i.e. $2m_{\text{e}}c^2$.

A photon can produce an electron–positron pair by interacting with an atomic electron rather than with a nucleus. The electron, being less massive than a nucleus, carries off more energy; the minimum photon energy for pair creation is correspondingly greater.

In the bubble chamber photograph (Fig. 6.3) γ-rays (whose tracks are not seen because they are uncharged) have entered from the left. In the event on the right of the picture a γ-ray photon has interacted with a nucleus of the bubble chamber liquid

Fig. 6.3
Electron–positron pair creation

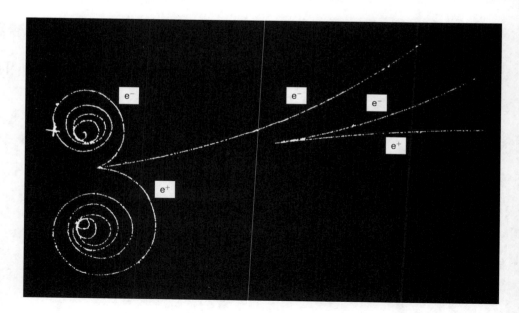

creating an electron–positron pair. A magnetic field directed out of the paper causes the particle tracks to curve in opposite directions. In the event on the left, a γ-ray has interacted with an atomic electron. The central track is that of the recoiling electron.

It can be shown (see section 9.7) that if two particles have the same charge, then the radii of their tracks are proportional to their momenta. Since all five of the tracks in Fig. 6.3 have been made by particles with the same rest mass, it follows from equation [6.6] that tracks of larger radius have been made by particles of greater energy. The reader should study the photograph with this in mind, noting that the interaction with the nucleus produces tracks of larger radius than the interaction with the electron, even though the γ-ray energy is the same in each case.

6.6 EXCHANGE PARTICLES

In classical physics the force between two charged particles is regarded as being due to the fact that each particle is in the electromagnetic field of the other. The quantum mechanical view is that the force is due to the exchange of photons which are emitted and absorbed by the particle, i.e. **the electromagnetic force (interaction) is transmitted by photons**. The photons exist for only a very short time ($\sim 10^{-20}$ s) and are known as **virtual photons** because they cannot be observed directly.

The reader may be wondering where the energy used to create these photons has come from. According to the **Heisenberg uncertainty principle**, the energy of a system that spends a time Δt in any particular state is uncertain to an extent ΔE, where

$$\Delta E \Delta t \approx h/2\pi$$

Thus a photon of energy ΔE can be created without violating the law of conservation of energy providing it exists for not more than a time Δt where $\Delta t \approx h/(2\pi \Delta E)$.

The electromagnetic interaction is one of the four fundamental interactions. The other three, the strong, weak and gravitational interactions, are also transmitted (or **mediated**) by exchange particles (Table 6.2). The particles (including the photon) are known as **gauge bosons**.

Table 6.2
The four fundamental interactions and their exchange particles

Interaction	Range	Relative strength	Typical decay time	Exchange particle(s)	Rest mass (GeV/c²)	Charge	Spin
Strong	$\sim 10^{-15}$ m	1	10^{23} s	Gluon	0	0	1
Electromagnetic	∞	10^{-2}	10^{20}–10^{-16} s	Photon	0	0	1
Weak	$\sim 10^{-18}$ m	10^{-5}	$\geq 10^{-10}$ s	W^+, W^-, Z^0	81, 81, 93	$+1, -1, 0$	1
Gravitational	∞	10^{-38}	—	Graviton*	0	0	2

*The graviton is predicted by theory but it has never been detected.

The weak interaction is an extremely short-range force (see Table 6.2). It is responsible for β-decay and for the decay of many unstable particles. **Processes which involve the weak interaction take place much more slowly than those involving either the electromagnetic interaction or the strong interaction.** For example, the average lifetime of a particle that decays by means of the weak interaction is about 10^{-10} s, whereas decays which proceed by way of the strong interaction and the electromagnetic interaction take place in about 10^{-23} s and 10^{-20} s to 10^{-16} s respectively.

Table 6.2 lists the relative strengths of the various interactions – note that **the stronger the interaction, the shorter the decay time**. The gravitational interaction is of no consequence in particle physics because the particles concerned have very small masses.

The strong interaction which binds nucleons together inside a nucleus is mediated by the exchange of virtual pions. We shall see later (section 6.11) that nucleons are not fundamental particles in the strict sense, but are composed of particles called **quarks**. At this more fundamental level the strong interaction is the force which binds the quarks, and is mediated by the exchange of **gluons**. The force which binds the nucleons is really only a remnant of the force which binds the quarks.

We have seen that a virtual particle can exist for a time Δt where

$$\Delta t \approx \frac{h}{2\pi\Delta E}$$

ΔE must be at least as big as the rest energy of the particle, and therefore for a particle whose rest mass is m_0, $\Delta E \approx m_0 c^2$, i.e.

$$\Delta t \approx \frac{h}{2\pi m_0 c^2}$$

If we assume that the speed of the particle is approximately equal to the speed of light, then its range, R, is given by

$$R \approx c\Delta t$$

i.e.
$$R \approx \frac{h}{2\pi m_0 c}$$

Thus **the range of an interaction is inversely proportional to the rest mass of the exchange particle.** It follows that the electromagnetic interaction and the gravitational interaction have infinite range because they are mediated by massless particles. The weak interaction and the force between nucleons, on the other hand, are short-range forces because they are due to the exchange of particles which have mass.

6.7 FEYNMAN DIAGRAMS

Feynman diagrams are used to simplify complex calculations in particle physics. A proper account of their usefulness is beyond the scope of this book, and we shall simply regard them as pictorial representations of particle interactions. Two examples are shown in Fig. 6.4. In Fig. 6.4(a) two electrons approach each other, exchange a virtual photon and repel each other as a result. Fig. 6.4(b) represents β^--decay. A neutron emits a virtual W^--boson and becomes a proton. The W^- then decays into an electron and an antineutrino.

Though Fig. 6.4(a) may appear to suggest otherwise, Feynman diagrams tell us nothing about the <u>paths</u> of the interacting particles – the angles at which the various lines meet have no <u>significance</u>.

Richard Feynman
(1918–88)

Fig. 6.4
Feynman diagrams of
(a) electron–electron
scattering, (b) β^--decay

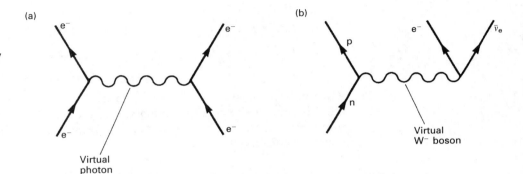

6.8 CLASSIFICATION OF PARTICLES

Particles other than the gauge bosons can be classified as being either **leptons** or **hadrons**.

Hadrons are particles that can take part in the strong interaction; **leptons** cannot.

Leptons

There are twelve leptons, six particle–antiparticle pairs (Table 6.3). **All leptons interact by way of the weak interaction; charged leptons can also interact through the electromagnetic interaction.***

L_e, L_μ and L_τ (Table 6.3) are called **lepton numbers**. Leptons obey a conservation law which requires that

Lepton numbers are conserved in all interactions.

*Leptons also interact by way of the gravitational interaction, but they are of such small mass that it is of no consequence.

Table 6.3
The Leptons

Name	Rest mass (MeV/c^2)	Spin	Particle					Antiparticle				
			Symbol	Charge	L_e	L_μ	L_τ	Symbol	Charge	L_e	L_μ	L_τ
Electron	0.511	$\frac{1}{2}$	e^-	-1	1	0	0	e^+	1	-1	0	0
Electron neutrino	Probably zero	$\frac{1}{2}$	ν_e	0	1	0	0	$\bar{\nu}_e$	0	-1	0	0
Muon*	105.7	$\frac{1}{2}$	μ^-	-1	0	1	0	μ^+	1	0	-1	0
Muon neutrino	Probably zero	$\frac{1}{2}$	ν_μ	0	0	1	0	$\bar{\nu}_\mu$	0	0	-1	0
Tau*	1784	$\frac{1}{2}$	τ^-	-1	0	0	1	τ^+	1	0	0	-1
Tau neutrino	Probably zero	$\frac{1}{2}$	ν_τ	0	0	0	1	$\bar{\nu}_\tau$	0	0	0	-1

*These are unstable with mean lifetimes of 2.2×10^{-6} s (muon) and 3×10^{-13} s (tau).
The muon is sometimes referred to as a μ meson – it should not be, the term used to be used but is now obsolete.

Consider, for example, the unstable particle μ^- which decays according to

$$\mu^- \rightarrow e^- + \nu_\mu + \bar{\nu}_e$$

$$L_e: \quad 0 \rightarrow 1 \quad 0 \quad -1 \qquad \text{i.e.} \quad 0 \rightarrow 0$$

$$L_\mu: \quad 1 \rightarrow 0 \quad 1 \quad 0 \qquad \text{i.e.} \quad 1 \rightarrow 1$$

$$L_\tau: \quad 0 \rightarrow 0 \quad 0 \quad 0 \qquad \text{i.e.} \quad 0 \rightarrow 0$$

The total values on the right-hand side are $L_e = 0$, $L_\mu = 1$ and $L_\tau = 0$, which are the same as the corresponding values on the left-hand side, i.e. each lepton number is conserved. Note also that the decay conforms to the general rule that charge is conserved.

Notes

(i) The lepton number of all hadrons and all gauge bosons is zero.

(ii) Particles have an intrinsic angular momentum and there is a quantum number, called the **spin quantum number** (or simply, **spin**), associated with this. Spin is a quantum mechanical concept – it has no classical analogue but it is sometimes helpful to think of a particle as a tiny sphere spinning about an axis through its centre.

(iii) The electron is stable because there is no lighter particle into which it can decay and still conserve charge.

> Particles which have half-integer spin ($\frac{1}{2}$, $\frac{3}{2}$, $\frac{5}{2}$, ...) are called **fermions**; those with zero or integer spin (0, 1, 2, ...) are called **bosons**. Fermions are subject to the **Pauli exclusion principle** (which states that two identical fermions in the same quantum mechanical system cannot have the same set of quantum numbers); bosons are not. **All leptons have spin = $\frac{1}{2}$ and therefore all leptons are fermions.**

Hadrons

Hadrons interact by way of the strong, the weak and the electromagnetic interactions.* They are divided into two groups – **baryons** and **mesons** (Table 6.4).

*Hadrons also interact by way of the gravitational interaction, but they are of such small mass that it is of no consequence.

Table 6.4
Properties of some hadrons

	Particle		Spin	Q	B	S	Rest mass (MeV/c²)	Mean lifetime (s)	Typical decay	Antiparticle
MESONS	Pion	π^0	0	0	0	0	135.0	0.8×10^{-16} *(1)	$\pi^0 \to \gamma + \gamma$	Self
		π^+	0	1	0	0	139.6	2.6×10^{-8}	$\pi^+ \to \mu + \nu_\mu$	π^-
	Kaon	K^+	0	1	0	1	493.7	1.2×10^{-8}	$K^+ \to \pi^+ + \pi^0$ or $\mu^+ + \nu_\mu$	K^-
		K^0	0	0	0	1	497.7	$\begin{cases} 8.9 \times 10^{-11} \\ 5.2 \times 10^{-8} \end{cases}$ *(2)	$K^0{}_S \to \pi^0 + \pi^0$ $K^0{}_L \to \pi^0 + \pi^0 + \pi^0$	$\overline{K^0}$
BARYONS	Proton	p	$\frac{1}{2}$	1	1	0	938.3	Stable		$\overline{p}$
	Neutron	n	$\frac{1}{2}$	0	1	0	939.6	898 *(3)	$n \to p + e^- + \overline{\nu}_e$	$\overline{n}$
	Lambda	Λ^0	$\frac{1}{2}$	0	1	-1	1116	2.6×10^{-10}	$\Lambda^0 \to p + \pi^-$ or $n + \pi^0$	$\overline{\Lambda^0}$
	Sigma	Σ^+	$\frac{1}{2}$	1	1	-1	1189	0.8×10^{-10}	$\Sigma^+ \to p + \pi^0$ or $n + \pi^+$	$\overline{\Sigma^+}$
		Σ^0	$\frac{1}{2}$	0	1	-1	1192	7.4×10^{-20} *(4)	$\Sigma^0 \to \Lambda^0 + \gamma$	$\overline{\Sigma^0}$
		Σ^-	$\frac{1}{2}$	-1	1	-1	1197	1.5×10^{-10}	$\Sigma^- \to n + \pi^-$	$\overline{\Sigma^-}$
	Omega	Ω^-	$\frac{3}{2}$	-1	1	-3	1672	0.8×10^{-10}	$\Omega^- \to \Lambda^0 + K^-$	Ω^+

*(1) π^0 decays by way of the electromagnetic interaction – hence the rapid decay.

*(2) K^0 can be regarded as a mixture of a short-lived state, K^0_S, and a longer-lived state, K^0_L, so that when K^0 decays two separate decays are observed.

*(3) Although free neutrons are unstable they can form stable nuclei when the energy released by the neutron decay is less than the energy with which they are bound into the nucleus.

*(4) Σ^0 is actually an excited state of Λ^0, hence the rapid decay by way of the electromagnetic interaction.

The table includes values of **baryon number** (B) and **strangeness** (S). These are quantum numbers that it has been found useful to assign to hadrons in order to account for the way they behave. (Leptons can be regarded as having $B = 0$ and $S = 0$.)

(i)   All baryons have $B = 1$, all antibaryons have $B = -1$ and all mesons, and antimesons have $B = 0$.

(ii)   Particles and their antiparticles have equal and opposite values of strangeness.

(iii)   All baryons (and antibaryons) are fermions; all mesons (and antimesons) are bosons.

Notes   (i)   The proton is the only baryon which is stable. All other baryons decay to protons, either directly (e.g. $\Sigma^+ \to p + \pi^0$) or indirectly (e.g. $\Omega^- \to \Lambda^0 + K^-$ followed by $\Lambda^0 \to p + \pi^-$). Mesons decay, either directly or indirectly, into photons or leptons.

(ii)   It is impossible to predict the actual lifetime of an individual particle – particles decay spontaneously some random interval after being created. The values listed in Table 6.4 are the mean lifetimes of particles of the type concerned. In the study of radioactivity half-life is used rather than mean lifetime. It can be shown that

$$\text{Half-life} = \log_e 2 \times \text{Mean lifetime}$$

6.9 STRANGENESS

The concept of strangeness was introduced in 1953 to account for the 'strange' behaviour of K mesons (kaons) and Λ and Σ baryons. These particles are produced by way of the strong interaction and therefore were also expected to decay by it, i.e. with lifetimes of the order of 10^{-23} s. Instead, they decay much more slowly, by way of the weak interaction with lifetimes of the order of 10^{-10} s. Furthermore, they are always produced in pairs — a kaon and a baryon together. This is known as **associated production**. Examples are:

$$p + \pi^- \rightarrow \Lambda^0 + K^0$$

and

$$p + \pi^- \rightarrow \Sigma^- + K^+$$

Bubble chamber photograph and explanatory diagram showing production and decay of the strange particles, K^0 and Λ^0. A π^- enters at the bottom and interacts with a proton in the bubble chamber liquid producing K^0 and Λ^0. Because they are both neutral, they do not leave tracks but betray their presence when they subsequently decay.
(A false-colour version of this photo appears on the back cover.)

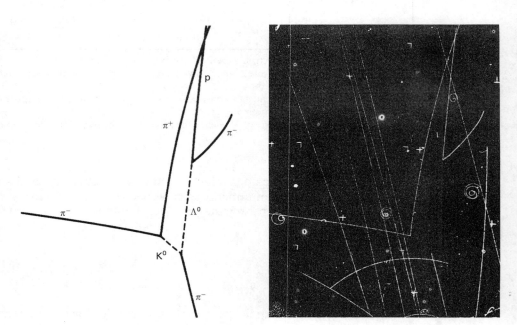

By introducing a quantum number (strangeness) that is conserved in strong interactions but not necessarily in weak ones (see section 6.10), the slower than expected decays and the phenomenon of associated production can be accounted for.

6.10 CONSERVATION LAWS

Particles obey the laws of conservation of momentum, angular momentum, mass–energy and charge. In addition:

(i)   Lepton number and baryon number are always conserved.

(ii)   Strangeness is conserved in all strong interactions and in all electromagnetic interactions. In weak interactions it changes by ± 1 or 0.

Production reactions (i.e. reactions in which two or more particles combine to form other particles) occur by way of the strong interaction and therefore strangeness is conserved in production reactions. Consider for example the production reaction

$$\pi^+ + p \rightarrow \Sigma^+ + K^+$$

Q:	1	$+1 \rightarrow$	1	$+1$	i.e.	$2 \rightarrow 2$
B:	0	$+1 \rightarrow$	1	$+0$	i.e.	$1 \rightarrow 1$
S:	0	$+0 \rightarrow$	-1	$+1$	i.e.	$0 \rightarrow 0$

Thus charge, baryon number and strangeness are all conserved.

When strange particles (i.e. particles with $S \neq 0$) <u>decay</u>, they do so by means of the weak interaction and strangeness changes by ± 1. Consider, for example, the strange particle Λ^0 which undergoes the weak decay

$$\Lambda^0 \rightarrow \pi^- + p$$

Q:	0	$\rightarrow -1$	$+1$		i.e.	$0 \rightarrow 0$
B:	1	$\rightarrow 0$	$+1$		i.e.	$1 \rightarrow 1$
S:	-1	$\rightarrow 0$	$+0$		i.e.	$-1 \rightarrow 0$

Thus charge and baryon number are conserved but strangeness increases by 1.

As a corollary, we observe that **any decay in which strangeness is not conserved must be a weak decay**.

Note The baryon numbers of the various particles are based on the proton being arbitrarily assigned a baryon number of $+1$. Values of strangeness are based on K^+ being assigned a strangeness of $+1$.

QUESTIONS 6C

1. Use Tables 6.3 and 6.4, and the conservation laws for charge, lepton number, baryon number and strangeness to determine which of the following reactions <u>cannot</u> occur.

(a) $\mu^+ \rightarrow e^+ + v_e$
(b) $\pi^+ \rightarrow \mu^+ + v_\mu$
(c) $\pi^+ + p \rightarrow \Sigma^+ + \pi^+$
(d) $\pi^- + p \rightarrow \Lambda^0 + K^+$
(e) $\pi^- + p \rightarrow \overline{\Sigma^+} + K^-$
(f) $\pi^0 \rightarrow e^- + \mu^+ + \overline{v}_e$
(g) $\Sigma^+ \rightarrow n + \pi^+$
(h) $K^+ + K^- \rightarrow \pi^0$
(i) $n \rightarrow p + e^- + v_e$
(j) $K^+ \rightarrow \mu^+ + v_\mu$
(k) $\Sigma^0 \rightarrow \pi^+ + \pi^-$
(l) $\pi^- + p \rightarrow n + \pi^0$
(m) $K^- \rightarrow \mu^- + \overline{v}_\mu$
(n) $K^- \rightarrow \pi^- + \pi^0$
(o) $p + p \rightarrow p + p + \pi^0$
(p) $p + p \rightarrow p + \Sigma^+ + \pi^0$

2. Use Table 6.4 and the conservation laws for charge, baryon number and strangeness to identify particle X in the following reactions
(a) $K^- + p \rightarrow K^+ + K^0 + X$
(b) $\pi^+ + n \rightarrow \Lambda^0 + X$
(c) $K^0 + p \rightarrow K^+ + X$

3. Give <u>two</u> reasons why a neutron cannot decay according to $n \rightarrow \pi^+ + e^-$

6.11 QUARKS

The leptons are thought to be genuinely fundamental particles; the hadrons are not. According to a proposal by Murray Gell-Mann$\star$ in 1964:

> **Hadrons** are composed of particles called **quarks** which cannot exist independently.

$\star$George Zweig made a similar proposal but the name 'quark' is due to Gell-Mann.

Quarks come in six **flavours**: up, down, strange, charmed, bottom and top. Initially we shall be concerned only with the first three of these. Some of their properties, together with those of the corresponding antiquarks, are listed in Table 6.5. Note that **each quark and its antiquark have equal and opposite values of charge, baryon number and strangeness**.

Table 6.5
Up, down and strange quarks

Quark	Charge (Q)	Baryon number (B)	Strangeness (S)	Spin
Down (d)	$-\frac{1}{3}$	$\frac{1}{3}$	0	$\frac{1}{2}$
Up (u)	$\frac{2}{3}$	$\frac{1}{3}$	0	$\frac{1}{2}$
Strange (s)	$-\frac{1}{3}$	$\frac{1}{3}$	$-1^\star$	$\frac{1}{2}$
Antidown ($\bar{\text{d}}$)	$\frac{1}{3}$	$-\frac{1}{3}$	0	$\frac{1}{2}$
Antiup ($\bar{\text{u}}$)	$-\frac{2}{3}$	$-\frac{1}{3}$	0	$\frac{1}{2}$
Antistrange ($\bar{\text{s}}$)	$\frac{1}{3}$	$-\frac{1}{3}$	1	$\frac{1}{2}$

*The strange quark has a strangeness of -1 (rather than $+1$) because strangeness quantum numbers were assigned to the various strange particles on an arbitrary basis before the existence of quarks was known.

Murray Gell-Mann
(b. 1929)

(i)   Baryons are composed of three quarks (qqq); antibaryons of three antiquarks ($\bar{\text{q}}\,\bar{\text{q}}\,\bar{\text{q}}$).

(ii)   Mesons are composed of a quark and an antiquark (q$\bar{\text{q}}$).

(iii) Quarks and antiquarks are fermions $\left(\text{spin} = \frac{1}{2}\right)$ and are therefore subject to the Pauli exclusion principle. They may combine with their spins parallel or antiparallel. This accounts for the fact that mesons have spin 0 ($\uparrow\downarrow$) or 1 ($\uparrow\uparrow$) and baryons have spin $\frac{1}{2}$ ($\uparrow\uparrow\downarrow$) or $\frac{3}{2}$ ($\uparrow\uparrow\uparrow$).

(iv)   The exchange particle that binds quarks together is the **gluon** – the mediator of the strong interaction.

The proton is composed of two up quarks and a down quark, i.e. the proton is uud. From table 6.5

$$\text{Charge } (Q) \quad = \tfrac{2}{3} + \tfrac{2}{3} - \tfrac{1}{3} = 1$$

$$\text{Baryon number } (B) = \tfrac{1}{3} + \tfrac{1}{3} + \tfrac{1}{3} = 1$$

$$\text{Strangeness } (S) \quad = 0 + 0 + 0 = 0$$

which are the values given in Table 6.4. The spins combine as ↑↑↓ to give spin $= \frac{1}{2} + \frac{1}{2} - \frac{1}{2} = \frac{1}{2}$ as required. The reader should confirm that the other combinations given in Table 6.6 are correct.

Table 6.6
Quark content of selected particles and antiparticles

Particle	Quark content	Antiparticle	Quark content
p	uud	$\bar{\text{p}}$	$\bar{\text{u}}\,\bar{\text{u}}\,\bar{\text{d}}$
n	udd	$\bar{\text{n}}$	$\bar{\text{u}}\,\bar{\text{d}}\,\bar{\text{d}}$
π^+	u$\bar{\text{d}}$	π^-	$\bar{\text{u}}$d
π^0	u$\bar{\text{u}}$,d$\bar{\text{d}}$★	π^0	u$\bar{\text{u}}$,d$\bar{\text{d}}$

★π^0 is actually a combination of u$\bar{\text{u}}$ and d$\bar{\text{d}}$; at this level it can be regarded as either u$\bar{\text{u}}$ or d$\bar{\text{d}}$.

Quarks feel all four of the fundamental interactions. In particular, the strong interaction binds quarks together as hadrons and the weak interaction between quarks is the basis of β-decay. In β^--decay, for example, a neutron decays by way of the weak interaction according to

$$\text{n} \rightarrow \text{p} + \text{e}^- + \bar{\nu}_\text{e}$$

What happens is that a d quark changes into a u quark by emitting a W$^-$ particle which subsequently decays into an electron and an antineutrino. The process is shown schematically and in terms of a Feynman diagram in Fig. 6.5.

Fig. 6.5
β^--decay in terms of the quark structure of neutrons and protons (a) schematically, (b) as a Feynman diagram

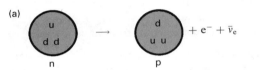
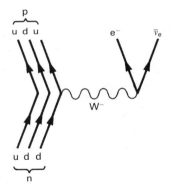

QUESTIONS 6D

1. Mesons and antimesons are q$\bar{\text{q}}$ combinations. Use Tables 6.4 and 6.5 to find the quark content of (a) K$^+$, (b) K^0. (c) Without further reference to either of the Tables, write down the quark structure of K$^-$, the antiparticle of K$^+$.

2. Baryons and antibaryons are qqq and $\bar{\text{q}}\bar{\text{q}}\bar{\text{q}}$ combinations respectively. Use Tables 6.4 and 6.5 to find the quark content of the following baryons and antibaryons: (a) Λ^0, (b) $\overline{\Lambda^0}$, (c) Σ^+, (d) $\overline{\Sigma^+}$, (e) Σ^-, (f) Σ^0, (g) Ω^-.

6.12 EXPERIMENTAL EVIDENCE FOR QUARKS

This was first provided by experiments at SLAC in 1969 in which high-energy (~ 10 GeV) electrons were fired at protons. The electrons had a de Broglie wavelength of the order of 10^{-16} m. This is about ten times smaller than the diameter of the proton and therefore the electrons were capable of resolving any structure the proton might have. If protons were structureless particles, their charge would be expected to be distributed uniformly throughout their volume and the electrons would be expected to pass through with very little deviation. It turned out that many electrons were scattered through large angles, indicating that they had interacted with charged, point-like objects inside the protons. Furthermore, these point-like objects had charges of $+\frac{2}{3}$ and $-\frac{1}{3}$, compelling evidence in favour of the quark model.

Notes (i) A major advantage of using electrons is that they do not feel the strong interaction – if they did, it would have been practically impossible to analyse the results.

(ii) The experiments are reminiscent of Rutherford's discovery of the nucleus through α-particle scattering experiments.

6.13 THE NEW QUARKS AND LEPTON–QUARK SYMMETRY

When Gell-Mann first suggested the existence of quarks in 1964, just three quarks (u, d and s) were sufficient to account for all the hadrons known at the time. By 1970, however, there was reason to believe that there must be a fourth quark.

One of the arguments in favour of this fourth quark was based on symmetry. At the time (apart from the gauge bosons) only seven truly fundamental particles (and the corresponding antiparticles) were known – four leptons (e, ν_e, μ and ν_μ) and three quarks (u, d and s). Since the electron neutrino is always associated with the electron, and the muon neutrino is always associated with the muon, the four leptons can be regarded as two doublets (e, ν_e) and (μ, ν_μ). The quarks could be regarded as a doublet (u, d) and a singlet (s). Sheldon Glashow and his co-workers suggested, on the grounds of symmetry, that there should be a fourth quark – the **charmed quark (c)** which would form a doublet with the strange quark, creating two quark doublets to match the two lepton doublets. They proposed that the charmed quark has spin $= \frac{1}{2}$, $Q = \frac{2}{3}$, $S = 0$ and a new quantum number, charm $(C) = 1$. **Charm, like strangeness, is conserved in strong and electromagnetic interactions but not necessarily in weak interactions**.

Sheldon Glashow
(b. 1932)

The existence of the charmed quark was confirmed in 1974. The SLAC and Brookhaven accelerator laboratories independently discovered a meson, now called J/ψ (J/psi) whose quark structure is $c\bar{c}$.

About a year after this **four lepton–four quark symmetry** had been established by the discovery of the charmed quark, it was broken by the discovery of a new lepton – **the tau lepton (τ)**. This implied that there should be an associated neutrino (the tau neutrino) and two more quarks (designated as top and bottom). The existence of one of these, the **bottom quark (b)**, was confirmed in 1977 by the discovery of the Υ (upsilon) meson. The bottom quark has a quantum number, **bottomness** $(B') = -1$. There is also evidence for the top quark but (as yet) not enough to confirm its existence beyond doubt. Properties of the six quarks are listed in Table 6.7. The corresponding antiquarks also have spin $= \frac{1}{2}$, but have opposite values of B, Q, S, C, B' and T. The properties of the leptons have been given previously – Table 6.3.

Table 6.7
Properties of the six quarks

	Mass (GeV/c^2)	Spin	Baryon number B	Charge Q	Strangeness S	Charm C	Bottomness B'	Topness T
d	0.01	$\frac{1}{2}$	$\frac{1}{3}$	$-\frac{1}{3}$	0	0	0	0
u	0.005	$\frac{1}{2}$	$\frac{1}{3}$	$\frac{2}{3}$	0	0	0	0
s	0.2	$\frac{1}{2}$	$\frac{1}{3}$	$-\frac{1}{3}$	-1	0	0	0
c	1.2	$\frac{1}{2}$	$\frac{1}{3}$	$\frac{2}{3}$	0	1	0	0
b	4.7	$\frac{1}{2}$	$\frac{1}{3}$	$-\frac{1}{3}$	0	0	-1	0
t	> 89	$\frac{1}{2}$	$\frac{1}{3}$	$\frac{2}{3}$	0	0	0	1

Antiquarks have spin $= \frac{1}{2}$ but opposite values of B, Q, S, C, B' and T.

The various leptons and quarks can be arranged in three sets known as **generations** with a lepton doublet and a quark doublet (and the corresponding antiparticles) in each generation. The first generation consists of the electron and its neutrino, together with the up and down quarks. All ordinary matter can be constructed from this generation alone. Current theories offer no explanation for the existence of the other two, apparently superfluous, generations.

6.14 THE ELECTROWEAK THEORY

According to this theory the electromagnetic interaction and the weak interaction are merely two different manifestations of a single interaction – the **electroweak interaction**. The theory regards the differences between the two interactions as being due to the difference in mass of the exchange particles involved, and predicts that at energies in excess of 100 GeV the two merge into a single interaction because the difference in mass of the exchange particles is no longer significant.

Perhaps the most notable success of the theory came with the discovery of the $W^{\pm}$ and Z^0 particles at CERN in 1982 and 1983. These were produced in high-energy collisions between protons and antiprotons, and were found to have masses within 1% of the values predicted on the basis of the electroweak theory. Many of its other predictions have also been confirmed and the theory is generally regarded as being correct.

The $W^{\pm}$ and Z^0 particles are the mediators of the weak interaction (see section 6.6). These, together with the mediator of the electromagnetic interaction, the photon, are the mediators of the electroweak interaction. $W^{\pm}$ and Z^0 are extremely heavy particles (see Table 6.2). The electroweak theory predicts the existence of an even heavier particle called the **Higgs boson**. So far, it has not been observed experimentally but experiments designed to find it are currently underway using the world's highest energy particle accelerators.

6.15 GRAND UNIFICATION THEORIES (GUTs)

These are attempts to unify the electroweak interaction with the strong interaction. In most GUTs these become one and the same interaction at energies in excess of 10^{15} GeV.

Many of the theories predict that the proton can decay into a positron and a neutral pion (in violation of conservation of both baryon number and lepton number) with a mean lifetime of 10^{29} to 10^{31} years. (This is a remarkably long time – the age of the Universe is about 10^{10} years!)

Many experiments are underway to find evidence of proton decay. One of these involves monitoring a large tank containing 8000 tonnes of water, and therefore over 10^{33} protons. An average lifetime of even 10^{31} years would imply a few hundred decays per year in this amount of water. None of the experiments has so far produced conclusive evidence of proton decay but the search continues.

The various GUTs are still very much unproven, but they are not totally without experimental support. In particular, they are able to account for the fact that the magnitude of the charge on the proton is the same as that on the electron.

CONSOLIDATION

Antiparticles are not constituents of ordinary matter.

Particles and their antiparticles have the same mass and the same average lifetime but opposite values of (any non-zero values of) charge, lepton number, baryon number, strangeness and charm.

The four fundamental interactions are transmitted by **exchange particles** known as **gauge bosons**.

Gauge bosons are <u>fundamental</u> particles in the strict sense.

The strong interaction binds neutrons and protons to form nuclei, and binds quarks (and antiquarks) to form hadrons.

The weak interaction is responsible for β-decay and the decay of many unstable particles. Processes which involve the weak interaction occur more slowly than those involving the electromagnetic interaction or the strong interaction.

Hadrons feel the strong interaction: leptons do not.

Leptons

Charged leptons feel electromagnetic, weak (and gravitational) interactions.

Neutral leptons (i.e. the various neutrinos) feel weak (and gravitational) interactions.

Leptons have spin $= \frac{1}{2}$ and therefore are fermions.

Leptons are <u>fundamental</u> particles.

Leptons have $L = 1$, antileptons have $L = -1$.

Hadrons

Hadrons feel all four interactions.

Hadrons are not fundamental particles; they are composed of quarks bound together by gluons.

Hadrons are of two types – **baryons** and **mesons** .

Mesons

All mesons are unstable

Mesons and antimesons have $B = 0$

$$\text{Mesons} = q\bar{q} \qquad \text{Antimesons} = q\bar{q}$$

Baryons

All baryons except the proton are unstable.

Baryons have $B = 1$; antibaryons have $B = -1$.

$$\text{Baryons} = qqq \qquad \text{Antibaryons} = \bar{q}\bar{q}\bar{q}$$

Strange particles (i.e. particles for which $(S \neq 0)$ are <u>produced</u> by way of the strong interaction and strangeness is conserved, but they <u>decay</u> by the weak interaction and strangeness changes by ± 1.

Conservation Laws

Charge, lepton number and baryon number are always conserved. Strangeness is conserved in all strong and electromagnetic interactions; it changes by ± 1 or 0 in weak interactions.

Quarks

Quarks are the constituents of hadrons and cannot exist independently.

Quarks are <u>fundamental</u> particles.

Quarks feel all four interactions.

Relativistic Equations

$$m = \frac{m_0}{\sqrt{1 - \dfrac{v^2}{c^2}}} \qquad p = mv \qquad p = \frac{Ev}{c^2}$$

$$E = mc^2 \qquad E^2 = m_0^2 c^4 + p^2 c^2$$

Relativistic KE $=$ Total energy $-$ Rest energy

$$\text{Lorentz factor} = \gamma = \frac{1}{\sqrt{1 - \dfrac{v^2}{c^2}}}$$

QUESTIONS ON CHAPTER 6

1. **(a)** Gamma photons have been observed to decay by pair production.
   (i) What is pair production?
   (ii) How can it be observed?
   (b) (i) Taking the mass of an electron as 9.1×10^{-31} kg, calculate the minimum energy equivalent of an electron-positron pair.
   (ii) Use your result from (i) to determine the maximum wavelength of gamma radiation for which pair production is possible.
   (The Planck constant, $h = 6.6 \times 10^{-34}$ J s, speed of light in vacuum $= 3.0 \times 10^8$ m s^{-1}.)   [O, '92]

2. In the process of pair creation a γ-ray photon ceases to exist and an electron–positron pair is created in its place.
   (a) Explain why this cannot happen unless the γ-ray interacts with some other particle.
   (b) Explain why the minimum photon energy for pair creation in which a γ-ray interacts with an electron is greater than that in which a γ-ray interacts with a proton.

3. State two differences between **(a)** leptons and hadrons, **(b)** mesons and baryons.

4. **(a)** The electromagnetic force and the gravitational force have a similarity in that they are both infinite in range. State *one* more similarity and *two* differences between these forces.
   (b) According to special relativity theory the inertial mass, m, of an electron moving with speed v is given by

   $$m = m_0 \left(1 - \frac{v^2}{c^2}\right)^{-1/2}$$

   where m_0 is the rest mass of the electron and c is the speed of light *in vacuo*.
   (i) Use the equation to explain what happens to the mass m if the electron is accelerated to speeds very close to that of light.
   How does the theory forbid electrons from travelling at speeds greater than c?
   (ii) Describe one additional relativistic effect exhibited by high-speed electrons.   [L, '91]

5. **(a)** Calculate the magnitude of the electrostatic force between two protons separated by 1×10^{-13} m.
   Without further calculation, write down the magnitude of the electrostatic force between two protons separated by only 1×10^{-14} m.
   (b) Two protons separated by a distance of 10^{-13} m accelerate away from each other, whereas protons within an atomic nucleus of diameter 10^{-14} m remain tightly bound together.
   From this statement, what conclusion can be drawn about the forces involved?
   In one or two sentences, say how this conclusion is related to the idea of particle exchange.
   ($e = 1.6 \times 10^{-19}$ C, $1/4\pi\varepsilon_0 = 9.0 \times 10^9$ m F^{-1}.)   [L (specimen), '92]

6.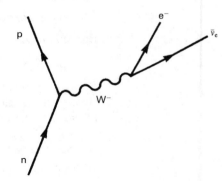

   The Feynman diagram above represents the β^--decay process.
   With reference to the diagram, describe each stage of the decay. Then redraw the diagram so that it shows the quark transformation that occurs in β^--decay.   [L (specimen), '92]

7. **(a)** Quantum theory shows that R, the range of an exchange particle, is given by

   $$R = \frac{h}{2\pi Mc}$$

   where
   h is the Planck constant
   ($= 6.63 \times 10^{-34}$ J s)
   c is the speed of light in vacuum ($= 3.00 \times 10^8$ m s^{-1}) and
   M is the mass of the exchange particle.

(i) Given that the range of the so-called *weak interaction* is 2.40×10^{-18} m, determine the ratio of the mass of a weak interaction exchange particle to the mass of a proton $(1.67 \times 10^{-27}$ kg).

(ii) What is the mass of the exchange particle responsible for *electromagnetic interactions*? Explain your answer.

(iii) State which of the *interactions*, (i) or (ii), has more similarity to that of gravity and give *two* reasons for your answer.

(b) β^--decay is now explained in terms of the weak interaction. State what β^--decay is and show how the weak interaction together with the quark structure of nucleons accounts for it. [L, '92]

8. A stationary neutron can decay as shown:
$$n \rightarrow p + e^- + \bar{v}$$
where $\bar{v}$ is an antineutrino.

(a) Determine the proton number and the nucleon number of an antineutrino and, in the light of your answers, state, giving your reasoning, *one* property you would expect of an antineutrino.

(b) Two varieties of quark are the up (u) quark, and the down (d) quark.
The quark structure of a proton is uud and that of a neutron is udd. What change in quark structure must occur when a neutron decays into a proton?
Which one of the four major interactions (forces) is responsible for this decay?
Given that a quark can only have a charge of $\pm \dfrac{e}{3}$ or $\pm \dfrac{2e}{3}$, determine the charges on the u and d quarks. [L, '91]

9. Use Tables 6.3 and 6.4 to identify particle X in each of the following reactions
(a) $p + \pi^- \rightarrow n + \pi^0 + \pi^- + X$
(b) $p + p \rightarrow \Lambda^0 + p + \pi^+ + X$
(c) $\tau^- \rightarrow \mu^- + v_\tau + X$
(d) $p + p \rightarrow \Lambda^0 + p + X$

10. Use the quark model to explain why the strangeness values of mesons and antimesons are restricted to 0 and ± 1.

11. The K^+ is a meson with strangeness $+1$. One of its common decay modes is $K^+ \rightarrow \pi^+ \pi^0$. Pions are not strange particles.

(a) Name the type of interaction responsible for the $K^+ \rightarrow \pi^+ \pi^0$ decay.

(b) The table below gives the properties of the six quarks. Deduce the possible quark content of the K^+, the π^+ and the π^0.

Type of quark	Charge	Baryon number	Strange-ness	Charm	Bottom-ness	Top-ness
u	$+\frac{2}{3}$	$\frac{1}{3}$	0	0	0	0
d	$-\frac{1}{3}$	$\frac{1}{3}$	0	0	0	0
c	$+\frac{2}{3}$	$\frac{1}{3}$	0	1	0	0
s	$-\frac{1}{3}$	$\frac{1}{3}$	-1	0	0	0
t	$+\frac{2}{3}$	$\frac{1}{3}$	0	0	0	1
b	$-\frac{1}{3}$	$\frac{1}{3}$	0	0	-1	0

(c) The rest mass of the proton is 938 MeV/c^2. The K^+ rest mass $= 0.53$ proton masses, and the π^+ rest mass $= \pi^0$ rest mass $= 0.15$ proton masses.
Assuming that a K^+ is stationary when it decays, show that the total energy of each pion produced in the decay is 249 MeV.
Hence calculate the momentum of each pion. Express your answer in units of MeV/c. [L (specimen), '92]

12. Tritium 3 can decay into helium 3 by the decay process shown below.
$$^3_1H \rightarrow ^3_2He + \beta^- + \bar{v}$$
Which of the four fundamental forces occurring in nature is responsible for
(a) the stability of electron orbits,
(b) the relative stability of nuclei, and
(c) the decay of tritium into helium.
Show in a table the relative strengths of, the ranges of, and the particles thought to be responsible for the mediation of, each of the four fundamental forces.
State why one of these four forces can be neglected in relation to (a), (b) and (c) above.
Which of the remaining three forces does this fourth force most closely resemble? Give *two* reasons for your choice. [L, '93]

13.

Type of quark	Charge	Baryon number	Strangeness
u	$+\frac{2}{3}$	$\frac{1}{3}$	0
d	$-\frac{1}{3}$	$\frac{1}{3}$	0
s	$-\frac{1}{3}$	$\frac{1}{3}$	-1

There are nine possible ways of combining u, d and s quarks and their associated antiquarks to make nine different mesons. List all the possible combinations. From your list select any strange mesons and state the charge and strangeness of each of these.

Three of the mesons in the list have zero charge and zero strangeness. What will distinguish these mesons from each other?

[L, '94]

14. A moving proton of kinetic energy 2.12 GeV collides with a stationary antiproton. These particles annihilate and a new particle of rest mass 2.74 GeV/c^2 and momentum 2.91 GeV/c is produced. Show that no other new particles have been produced as a result of this collision.

(Rest mass of proton = 0.938 GeV/c^2.)

[L, '94]

15. Strangeness, S, is a property of some elementary particles. Listed opposite are six particles each with their strangeness, S, and charge, Q.

Particle	Q	S
K⁻	−1	−1
K⁺	+1	+1
K⁰	0	+1
Λ	0	−1
n	0	0
p	+1	0

Find the charge and strangeness of X in the following interaction, given that strangeness is conserved.

$$K^- + p \rightarrow K^0 + K^+ + X$$

What do you understand by the term *conserved* in this example?

[L, '94]

16. A proton and an antiproton of equal energy collide and annihilate each other. Calculate the minimum energy released from the annihilation and the wavelengths of the photons emitted. Why must more than one photon be emitted?

(Speed of light in vacuum, $c = 3.0 \times 10^8$ m s⁻¹, mass of proton = mass of antiproton = 1.67×10^{-27} kg, the Planck constant, $h = 6.63 \times 10^{-34}$ J s.)

[L, '94]

7

PARTICLE PHYSICS AND COSMOLOGY

7.1 EXPANSION OF THE UNIVERSE

If an observer and a source of light are in relative motion, the wavelength of the light as measured by the observer is different from the actual wavelength of the light – a phenomenon known as the **Doppler effect**. When the source and the observer are moving away from each other the shift is to longer wavelengths – the light is said to be **red shifted**. When the source and the observer are approaching each other the shift is to shorter wavelengths, i.e. **violet shifted**. In both cases the extent of the shift increases as the relative velocity of the source and the observer increases.

When the light emitted by a star is examined spectroscopically, it is found that each line in the spectrum of any particular element in the star occurs at a different wavelength from that of the corresponding line in the spectrum of the same element in the laboratory. The shifts in wavelength are interpreted as being due to the Doppler effect.

The galaxies, apart from a few which are close to us, all exhibit a red shift and are therefore moving away from the Earth. In 1929 an American astronomer, Edwin Hubble, made extensive measurements of the galactic red shifts. Interpreting them as being due to the Doppler effect he found that

> The speed at which a galaxy is receding from us is proportional to its distance from us (**Hubble's law**).

Edwin Powell Hubble
(1889–1953)

This should not be taken to imply that we occupy some unique place in the Universe – the galaxies are not only moving away from us, they are also moving away from each other. If Hubble had been able to make his measurements from any other galaxy, he would have obtained an equivalent result.

7.2 THE BIG BANG

According to Hubble's law, the galaxies are receding at speeds proportional to their distances from us, i.e. those that are farthest away are moving the fastest. The most commonly accepted explanation of this is that at some time in the past the Universe was in a state of extremely high concentration and that as a result of some gigantic explosion (**the Big Bang**) it has been expanding ever since.

Hubble's law can be expressed as

$$v = H_0 d \qquad\qquad [7.1]$$

where v is the velocity at which a galaxy a distance d away from us is receding, and H_0 is a constant known as **Hubble's constant**. It follows from equation [7.1] that matter a distance d away from us has been travelling for a time t where

$$t = \frac{d}{v} = \frac{1}{H_0}$$

Since this matter is supposed to have been moving since the Universe began,

$$\text{Age of Universe} = \frac{1}{H_0}$$

We can expect the actual age to be somewhat less than this because the galaxies will have been slowing down as a result of their mutual gravitational attraction. Furthermore, there is some uncertainty in the value of H_0 because of uncertainties in the estimated distances of the more distant galaxies. When these various factors are taken into account the age of the Universe is estimated to be between ten and twenty billion (10^9) years.

Note   The reader should not take what we have said to mean that all the matter in the Universe was once concentrated at a single point in the Universe, but that that point was the Universe. The Big Bang created an expansion of space itself, it did not cause matter to rush outwards to occupy an already existing space. This is not an easy concept; it might help to consider two points, A and B, on the surface of a balloon (Fig. 7.1(a)). If we inflate the balloon (Fig. 7.1(b)), A and B become farther apart because the space they occupy, the surface of the balloon, has expanded. They have not moved apart by moving through an already existing space as in Fig. 7.1(c).

Fig. 7.1
To illustrate the difference between the expansion of space and matter moving through an already existing space

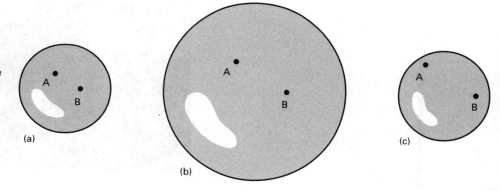

(a)

(b)

(c)

7.3 EVIDENCE FOR THE BIG BANG MODEL OF THE UNIVERSE

The Big Bang model is able to account for the expansion of the Universe. It has had two other notable successes.

(i) It predicts that the ratio, by mass, of $_1^1$H to $_2^4$He should be 3 to 1 and this has been confirmed by observation.

(ii) It is able to account for the so-called **microwave background radiation** – the radiation which is characteristic of a black body* at a temperature of 2.7 K and which appears to be uniformly distributed throughout the Universe. It was discovered by AA Penzias and RW Wilson in 1964. The Big Bang theory takes it to be a remnant of the early Universe and had predicted its existence.

It is these two successes, rather than its ability to account for the expansion, that has made the Big Bang model the most widely accepted theory of the Universe. The idea that the Universe is expanding is based on Hubble's law, which is itself based on the idea that the galactic red shifts are due to the Doppler effect. Some cosmologists regard the red shifts as being due to a gravitational effect rather than to the Doppler effect. If they are correct and if the Big Bang model did no more than account for the expansion, it is unlikely that it would have the widespread support that it has.

7.4 THE EVOLUTION OF THE UNIVERSE

As we have seen, the Universe is believed to have originated in an immense explosion (the Big Bang) between ten and twenty billion years ago (i.e. between 1×10^{10} and 2×10^{10} years ago). At that stage the Universe was extremely dense and extremely hot. It has been expanding ever since, increasing its total gravitational potential energy. As a consequence, there has been a corresponding decrease in the average kinetic energy of the particles making up the Universe, and therefore a corresponding decrease in its temperature.

According to the various grand unified theories, all four of the basic interactions would have been unified as a single interaction when the Universe was first formed. As it cooled, and the average particle energy decreased, first the gravitational and then the strong interaction would begin to act as separate interactions. At even lower energies, the weak interaction would become distinct from the electro-magnetic interaction.

As the Universe cooled, quarks combined to form neutrons and protons, and these subsequently combined to form nuclei. Finally, nuclei joined with electrons to form atoms.

Atoms could not exist whilst particle energies were higher than the energy required to remove electrons from atoms. This, of course, depends on the strength of the electromagnetic interaction. Nuclei could form when the Universe was much hotter, simply because it is the (much stronger) strong interaction that is responsible for binding nucleons together.

A very much simplified account of how the Universe is thought to have evolved as it cooled is given in Table 7.1.

*A black body is a perfect emitter and absorber of radiation. See, for example, *A-Level Physics* by R. Muncaster (Stanley Thornes).

Table 7.1
Evolution of the cooling
Universe

Time since Big Bang, Average energy per particle	Main features
$< 10^{-43}$ s, $> 10^{19}$ GeV	Current theories do not extend into this era, but it is supposed that all four interactions were unified.
10^{-43} s, 10^{19} GeV	Gravity now separate but strong interaction still coupled to electroweak, therefore no distinction between quarks and leptons.
10^{-35} s, 10^{15} GeV	**Quark–Lepton Era** Strong and electroweak interactions become uncoupled. Quarks and leptons now distinct. **Universe is now a mixture of quarks, leptons and the various exchange particles.** Electromagnetic and weak interactions become uncoupled at 100 GeV.
10^{-5} s, 1 GeV	**Hadron Era** Strong force begins to bind quarks to form **hadrons** (n, p, $\bar{\text{n}}$, $\bar{\text{p}}$ and various mesons and antimesons).
100 s, 0.1 MeV	**Plasma Era** Neutrons and protons combine to form **deuterons ($^2_1$H) followed by other light nuclei** according to $$^2_1\text{H} + {}^1_0\text{n} \rightarrow {}^3_1\text{H} \qquad \text{and} \qquad {}^2_1\text{H} + {}^1_1\text{p} \rightarrow {}^3_2\text{He}$$ then $$^3_1\text{H} + {}^1_1\text{p} \rightarrow {}^4_2\text{He} \qquad \text{and} \qquad {}^3_2\text{He} + {}^1_0\text{n} \rightarrow {}^4_2\text{He}$$ <u>Small</u> amounts of $^7_3$Li and $^7_4$Be are also produced, after which the building of nuclei very nearly ceases because there are no stable nuclei with mass numbers of either 5 or 8. Universe is now a high-temperature ($\sim 10^9$ K) **plasma** consisting mainly of protons, helium nuclei, electrons, photons and neutrinos.
700 000 years, 0.1 eV	**Atomic Era** Electrons become attached to nuclei to form **atoms**. No nuclei–nuclei repulsion and therefore galaxies, and then stars, form under the effect of gravity. Thermonuclear fusion occurs in stars, initially binding triplets of $^4_2$He to produce $^{12}_6$C and then nuclides of higher mass number up to $^{56}_{26}$Fe. Heavier nuclei are produced by neutron capture followed by β^--decay.

7.5 OPEN AND CLOSED UNIVERSES

It is assumed that the general expansion of the Universe is slowing down as a result of gravitational attraction. If this attraction is strong enough, the expansion will eventually stop and the Universe will begin to contract, ending in what is known as **'the Big Crunch'**. Such a universe is called a **closed universe**; one which expands indefinitely is an **open universe**.

Whether the Universe is open or closed depends on its average density. If this exceeds a value known as the **critical density** (calculated to be 6×10^{-27} kg m^{-3}), the Universe will eventually contract. If the average density is less than this value, on the other hand, the expansion will continue for ever – becoming ever slower, but never stopping. The various possibilities are shown graphically in Fig. 7.2. Note that if Hubble's law were obeyed <u>exactly</u> (which is not believed to be the case), the Universe would expand for ever <u>at a constant rate</u>. We do not know which of the other three curves we are on. If we are on the lower curve, we must be left of centre because the evidence is that the Universe is not contracting at the present time.

Fig. 7.2
Open and closed
universes compared

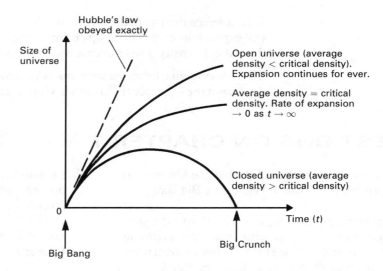

Estimates of the average density of the Universe, based on estimates of the total mass of all known material (e.g. stars, intersteller gas, black holes and the mass equivalent of radiation) give a value which is only a few per cent of the critical density. This implies that the Universe is open and will expand for ever. However, observations of the motions of galactic clusters suggest that the Universe contains a large amount of matter that is otherwise undetectable. This is known as **dark matter** and its nature can only be guessed at. One suggestion is that it consists of particles called **photinos**, the fermion equivalent of the photon predicted by supersymmetry theories. Another possibility is that neutrinos have non-zero rest mass. Neutrinos are thought to be so numerous that a rest mass of even as little as $30\,\text{eV}/c^2$ would be sufficient to stop the expansion.

CONSOLIDATION

Light emitted by galaxies is red shifted. If this is due to the Doppler effect, the galaxies must be moving away from us (and from each other).

Hubble's law

The speed at which a galaxy is receding is proportional to its distance away, i.e.

$$v = H_0 d$$

Hubble's law implies that the Universe was once in a state of extremely high concentration and that it has been expanding ever since.

The Big Bang model accounts for

(i)   the expansion of the Universe,

(ii)   the observed ratio of ^1_1H to ^4_2He, and

(iii)   the 2.7 K microwave background radiation.

The Universe cooled as it expanded and quarks combined to form neutrons and protons. These subsequently formed nuclei which eventually joined with electrons to form atoms.

The rate at which the Universe is expanding is decreasing because of gravitational effects.

If the average density of the Universe is less than the **critical density**, the Universe will expand for ever (i.e. an **open universe**). If the average density is greater than the critical density, the Universe will eventually contract (i.e. a **closed universe**).

There is evidence from the motions of galactic clusters that the Universe contains **dark matter** – matter that is otherwise undetectable.

QUESTIONS ON CHAPTER 7

1. What evidence is there **(a)** that the Universe is expanding, **(b)** that it began as a Big Bang?

2. Spectroscopic investigations of a distant galaxy reveal that it is receding from the Earth at $7.4 \times 10^7 \, \text{m s}^{-1}$. Use Hubble's law to determine the distance of the galaxy from the Earth
   (a) in metres
   (b) in light-years.
      (Hubble's constant $= 2.4 \times 10^{-18} \, \text{s}^{-1}$, speed of light $= 3.0 \times 10^8 \, \text{m s}^{-1}$.)

3. The Universe is said to be expanding. The diagram shows a deflated balloon with three points A, B and C marked on it.

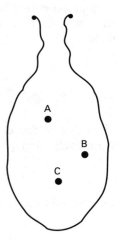

   Draw labelled diagrams showing what the balloon would look like when it is
   (a) half inflated, and
   (b) fully inflated,
   and use your diagrams to explain the concept of an expanding universe.   [L, '91]

4. Explain briefly how studies of line spectra from distant galaxies lead to the conclusion that the Universe is expanding.
   Cosmologists often use diagrams similar to that below when describing the expansion of the Universe.

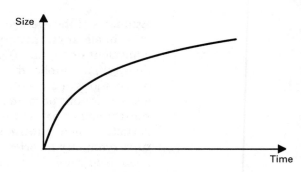

   In two or three sentences, describe the way in which the expansion rate changes as the Universe evolves and state the reason for this change.
   In the diagram above the Universe is assumed to have an average mass density exactly equal to the 'critical density'.
   The matter so far observed in the Universe suggests an average density below the critical value. Copy the above diagram and on it show how the Universe would evolve if this were the case.
   It is possible that the neutrino has a finite small mass. It is thought that neutrinos are very abundant in the Universe, and that their mass may be sufficient to raise the average density above the critical value. On your copy of the diagram, show how the Universe would evolve in this case.
   On your diagram label the 'Big Bang' and the 'Big Crunch'.   [L (specimen), '92]

8

PARTICLE ACCELERATORS

8.1 DEFLECTION OF CHARGED PARTICLES IN ELECTRIC AND MAGNETIC FIELDS

Electric Fields

A particle of charge Q in a field of intensity E is subject to a force F given by

$$F = QE$$

The force on <u>positively</u> charged particles is in the <u>same</u> direction as the field; that on <u>negative</u> particles is in the <u>opposite</u> direction to the field.

Consider a positively charged particle of charge Q moving with velocity v entering a <u>uniform</u> field of intensity E which is perpendicular to its direction of motion (Fig. 8.1). Once in the field the particle is subject to a force QE in the positive direction of the y-axis, and therefore by Newton's second law acquires an acceleration QE/m in this direction, where m is the mass of the particle. At the instant it enters the field its y-component of velocity is zero, and therefore after a time t in the field it will have undergone a vertical displacement, y, given by $s = ut + \frac{1}{2}at^2$ as

$$y = 0 + \tfrac{1}{2}\left(\frac{QE}{m}\right)t^2$$

i.e. $\qquad y = \tfrac{1}{2}\left(\frac{QE}{m}\right)t^2 \qquad\qquad\qquad\qquad\qquad [8.1]$

Fig. 8.1
Deflection of a positively
charged particle in an
electric field

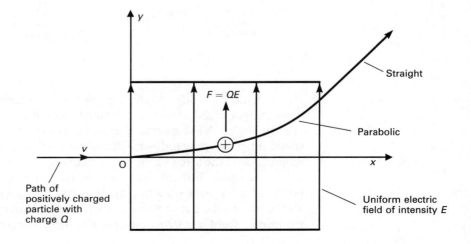

The x-component of velocity is unaffected by the field and therefore x, the particle's horizontal displacement from O, is given by

$$x = vt \qquad\qquad\qquad [8.2]$$

Eliminating t between equations [8.1] and [8.2] gives

$$y = \left(\frac{QE}{2mv^2}\right)x^2$$

This is the equation of a parabola, i.e. **the path of a charged particle whilst in a uniform electric field is parabolic**. Once the particle has left the field it travels in a straight line. Since the particle has gained a y-component of velocity whilst in the field and there has been no change in the x-component, <u>its kinetic energy has increased.</u>

Magnetic Fields

A particle of charge Q moving with velocity v at right angles to a magnetic field of flux density B experiences a force F given by

$$F = BQv$$

The force is perpendicular to both the field direction and the velocity, and its direction is given by Fleming's left-hand rule.

Consider a positively charged particle of charge Q moving with velocity v into a <u>uniform</u> magnetic field of flux density B which is at right angles to its direction of motion (Fig. 8.2).

Fig. 8.2
Deflection of a positively charged particle in a magnetic field

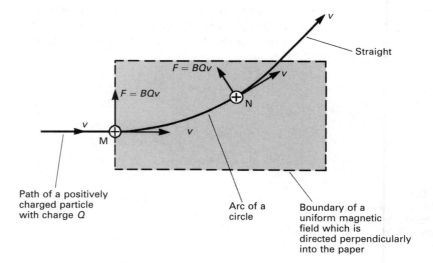

Path of a positively charged particle with charge Q

Arc of a circle

Boundary of a uniform magnetic field which is directed perpendicularly into the paper

On entering the field at M the particle feels a force F as shown and is deflected. (A negatively charged particle would be deflected in the opposite sense.) The force is at right angles to the direction of motion of the particle and therefore can neither speed it up nor slow it down. When the particle reaches some other point N, the <u>magnitude</u> of the force acting on it is the same as it was at M (since none of B, Q and v has changed) but the direction of the force is different. Thus, the force is perpendicular to the direction of motion at all times and has a constant magnitude, and therefore the particle travels with constant speed along a <u>circular arc</u>. Thus, **a magnetic field does not change the kinetic energy of a charged particle**.

The radius of curvature, r, of the path is given by Newton's second law as

$$BQv = \frac{mv^2}{r}$$ [8.3]

i.e. $$r = \frac{mv}{BQ}$$

Crossed Fields

If a uniform electric field and a uniform magnetic field are perpendicular to each other in such a way that they produce deflections in opposite senses, they are known as **crossed fields**. If the forces exerted by each field are of the same size, then

$$BQv = QE$$

i.e. $$v = E/B$$

Fig. 8.3
Principle of a velocity selector

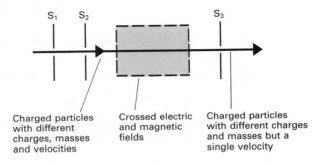

Charged particles with different charges, masses and velocities

Crossed electric and magnetic fields

Charged particles with different charges and masses but a single velocity

Fig. 8.3 shows how crossed fields can be used as a **velocity selector**, i.e. to select charged particles of a single velocity from a beam containing particles with a range of different velocities. Slits S_1 and S_2 confine the particles to a narrow beam. The only particles which are undeflected, and therefore which emerge from slit S_3, are those whose velocity v is given by $v = E/B$.

EXAMPLE 8.1

Refer to Fig. 8.4. A beam of electrons is accelerated through a PD of 500 V and then enters a uniform electric field of strength 3.00×10^3 V m^{-1} created by two parallel plates each of length 2.00×10^{-2} m. Calculate: (a) the speed, v, of the electrons as they enter the field, (b) the time, t, that each electron spends in the field, (c) the angle, θ, through which the electrons have been deflected by the time they emerge from the field. (Specific charge (e/m) for electron $= 1.76 \times 10^{11}$ C kg^{-1}.)

Fig. 8.4
Diagram for Example 8.1

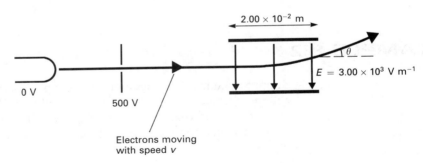

2.00×10^{-2} m

θ

$E = 3.00 \times 10^3$ V m^{-1}

0 V

500 V

Electrons moving with speed v

Solution

(a) The kinetic energy gained by an electron is equal to the work done by the PD. Therefore

$$\tfrac{1}{2}mv^2 = eV$$

i.e. $v = \sqrt{\dfrac{2eV}{m}}$

$$= \sqrt{2 \times 1.76 \times 10^{11} \times 500} = 1.327 \times 10^7$$

i.e. Speed on entering field $= 1.33 \times 10^7 \, \mathrm{m\,s^{-1}}$

(b) The horizontal component of velocity is unaffected by the field between the plates and is therefore constant. It follows that t is given by

$$t = \frac{2.00 \times 10^{-2}}{1.327 \times 10^7} = 1.507 \times 10^{-9}$$

i.e. Time between plates $= 1.51 \times 10^{-9} \, \mathrm{s}$

(c) The electric field, E, exerts a force, F, on each electron where

$$F = eE$$

By Newton's second law this gives each electron an acceleration, a, towards the top of the page, where

$$a = \frac{F}{m} = \frac{eE}{m}$$

On emerging from the field the electron will have gained a vertical component of velocity, v_y, given by $v = u + at$ as

$$v_y = 0 + \left(\frac{eE}{m}\right)t$$

i.e. $v_y = \left(\dfrac{e}{m}\right) \times E \times t$

$$= 1.76 \times 10^{11} \times 3.00 \times 10^3 \times 1.507 \times 10^{-9} = 7.957 \times 10^5$$

Since $\tan\theta = v_y/v$

$$\tan\theta = \frac{7.957 \times 10^5}{1.327 \times 10^7} = 5.996 \times 10^{-2}$$

$\therefore \quad \theta = 3.4°$

EXAMPLE 8.2

Suppose that in an arrangement of the type described in Example 8.1, particles of charge, Q, and mass, M, are accelerated by a PD, V, and then enter a field of strength, E, between plates of length, d. Obtain an expression for the angle, θ, through which the particles will have been deflected by the time they leave the plates.

Solution

The speed, v, on entering the field between the plates is given by

$$\tfrac{1}{2}Mv^2 = QV \qquad \therefore \qquad v = \sqrt{\frac{2QV}{M}}$$

The time, t, between the plates is given by

$$t = \frac{d}{v}$$

The vertical acceleration, a, is given by

$$a = \frac{QE}{M}$$

The vertical component of velocity, v_y, is given by $v = u + at$ as

$$v_y = \left(\frac{QE}{M}\right)\frac{d}{v}$$

$$\therefore \qquad \tan\theta = \frac{v_y}{v} = \frac{QEd}{Mv^2}$$

Substituting for v gives

$$\tan\theta = \frac{QEd}{M}\frac{M}{2QV} = \frac{Ed}{2V}$$

i.e. $\qquad \theta = \tan^{-1}\left(\frac{Ed}{2V}\right)$

Note that θ depends only on E, d and V – it does not depend on either Q or M. This result is interesting in itself, but it also serves to illustrate that we could have obtained the answer to Example 8.1 even if we had not been given the value of the specific charge of the particles involved. It further illustrates the effort that can be saved by not putting in numerical values until it is absolutely necessary!

QUESTIONS 8A

1. Calculate the speed of a proton which has been accelerated through a PD of 400 V.
   (Mass of proton = 1.67×10^{-27} kg, charge on proton = 1.60×10^{-19} C.)

2. An electron is moving in a circular path at 3.0×10^6 m s^{-1} in a uniform magnetic field of flux density 2.0×10^{-4} T. Find the radius of the path.
   (Mass of electron = 9.1×10^{-31} kg, charge on electron = 1.6×10^{-19} C.)

8.2 THE VAN DE GRAAFF ACCELERATOR

This device (Fig. 8.5) uses the high PD produced by a Van de Graaff generator to accelerate charged particles. The PD can be kept constant to within ±0.01% and therefore particles of very uniform energy can be produced.

The belt is driven at high speed past a series of metal points at X. The points are at a high positive potential with respect to earth and the insulation of the gas around them breaks down. Positively charged ions are 'sprayed' on to the belt and these are carried up into the collecting sphere. This induces a negative charge on the

Fig. 8.5
The Van de Graaff
accelerator

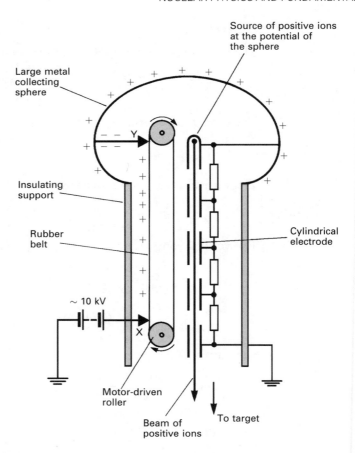

inner surface of the sphere and a corresponding positive charge on its outer surface. The sphere is in electrical contact with a second set of metal points, and the induced negative charge is sprayed from these on to the belt at Y. This neutralizes the positive charge on the belt so that the whole process can be repeated. In this way a considerable amount of positive charge builds up on the outer surface of the collecting sphere.

The maximum PD attainable ($\sim 10\,\mathrm{MV}$) depends on the insulating properties of the gas around the sphere. It is increased by enclosing the whole device in a pressure vessel containing an inert gas at a pressure of about 20 atmospheres.

An ion-source, at the same potential as the sphere, produces ions which enter a column of cylindrical electrodes, each of which is at a lower potential than the one above it. The ions are accelerated as they pass through the gaps between the electrodes. The non-uniform electric fields in the gaps have the effect of focusing the particles into a fine beam.

8.3 THE CYCLOTRON

This consists of two semicircular boxes (called **dees**) separated by a small gap and contained in an evacuated chamber in the uniform magnetic field provided by a large electromagnet (Fig. 8.6(a)). Charged particles (e.g. protons) produced by an ion-source at the centre of the device enter one or other of the dees and move along a circular path under the influence of the magnetic field (Fig. 8.6(b)).

Fig. 8.6
The cyclotron (a) side
view, (b) top view

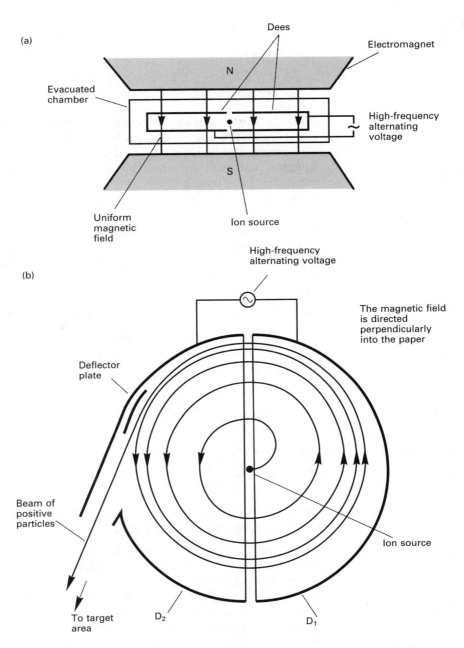

A high-frequency alternating PD is connected between the dees. Its frequency is such that its polarity reverses at the same rate as the particles cross from one dee to the other. It follows that any particle which is accelerated across the gap between the dees will be accelerated <u>every</u> time it crosses from one dee to the other.

If a particle of charge Q enters D_1 (say) with speed v, it will move on a semicircular path of radius r where

$$BQv = \frac{mv^2}{r} \qquad \text{(equation [8.3])}$$

i.e. $\qquad r = \dfrac{mv}{BQ}$ [8.4]

where B is the flux density of the magnetic field. If the particle takes a time t to travel through D_1 (a distance πr), then

$$t = \frac{\pi r}{v}$$

Therefore by equation [8.4]

$$t = \frac{\pi m}{BQ} \qquad\qquad [8.5]$$

Thus **the time to travel around the semicircle is constant provided B is constant and, in particular, is independent of both v and r.** On completing the semicircular path the particle will be accelerated across the gap to D_2 and, by equation [8.4], will then travel along a path of larger radius because its speed has increased. It follows from equation [8.5] that it will take the same time to travel round D_2 as it did around D_1. Therefore by the time it leaves D_2 the polarity will have reversed again and the particle will be accelerated across the gap into D_1. Thus the speed of the particle is increased every time it crosses between the dees and it will travel along a path of increased radius each time. It therefore spirals outwards going ever faster. On reaching the maximum radius the particles are deflected by a charged plate (negatively charged in the case of positive particles) and directed towards the target area.

It follows from equation [8.4] that the energy, E, of the particles is given by

$$E = \frac{1}{2}\,mv^2 = \frac{1}{2}\,m\left(\frac{BQr}{m}\right)^2 \qquad\qquad [8.6]$$

The maximum energy, E_m, corresponds to the maximum radius, r_m, and is given by equation [8.6] as

$$E_m = \frac{1}{2}\,m\left(\frac{BQr_m}{m}\right)^2$$

i.e.
$$E_m = \frac{1}{2}\,\frac{B^2Q^2r_m^2}{m}$$

Thus high particle energies require large B and large r_m, i.e. strong magnetic fields which are uniform over a large area.

Cyclotrons are used to accelerate heavy particles such as protons and deuterons rather than electrons. This is because the relativistic increase in mass which occurs with increasing velocity causes the particles to orbit ever more slowly (see equation [8.5]). Eventually, they cease to be in phase with the alternating PD between the dees. Protons can be accelerated to higher energies than electrons simply because for any given energy they move at a much lower speed. The upper limit for protons is about 25 MeV.

Notes   (i)   The polarity of an alternating PD reverses twice per cycle. Since the PD between the dees reverses twice per complete orbit, it follows that its period, T, is equal to the orbital period of the particles and therefore by equation [8.5]

$$T = 2t = \frac{2\pi m}{BQ}$$

The frequency, f, of the alternating PD is $1/T$ and therefore

$$f = \frac{BQ}{2\pi m} \qquad \textbf{(the cyclotron frequency)}$$

In order to achieve this condition, known as **cyclotron resonance**, either f or B is adjusted.

(ii) Each dee is at a uniform (albeit time-varying) potential and therefore the particles are not affected by the <u>electric</u> field whilst they are inside either of the dees.

(iii) The energy with which the particles emerge does not depend on the size of the voltage between the dees. However, it is desirable that the particles do not orbit more than about 50 times because the greater the total path length, the poorer the focusing. It follows that large dee voltages are required.

QUESTIONS 8B

1. A particular cyclotron has a maximum path radius of 0.47 m and a magnetic field of 1.6 T. What is the maximum particle energy (in MeV) that the device can produce for (a) protons, (b) deuterons, (c) α-particles? You may assume that a deuteron has twice the mass of a proton and that an α-particle has four times the mass of a proton.
($e = 1.60 \times 10^{-19}$ C, mass of proton $= 1.67 \times 10^{-27}$ kg.)

2. A cyclotron produces protons with an energy of 25 MeV and employs a dee voltage of 250 kV. How many orbits do the protons make?

3. Protons in a cyclotron operating at 16.0 MHz have a maximum path radius of 40.0 cm. Find (a) the flux density of the magnetic field, (b) the energy (in MeV) with which the protons emerge.
($e = 1.60 \times 10^{-19}$ C, mass of proton $= 1.67 \times 10^{-27}$ kg.)

4. A cyclotron produces protons with a kinetic energy of 16 MeV. The voltage between the dees is 100 kV at a frequency of 25 MHz. What is the <u>total</u> time that each proton spends in orbit?

8.4 THE LINEAR ACCELERATOR (LINAC)

A linear accelerator (LINAC) accelerates charged particles (usually protons or electrons) in a <u>straight</u> line along the axis of an evacuated tube. They are of two main types: **drift tube accelerators**, which are normally used to accelerate protons, and **travelling-wave accelerators**, which are used for electrons.

The Stanford Linear Accelerator Center (SLAC) in California showing the 2-mile long linear accelerator

Drift Tube Accelerators

In the drift tube accelerator (Fig. 8.7) A, B, C . . . are hollow metal cylinders (called **drift tubes**) of progressively greater length. They are connected alternately to opposite terminals of a high-frequency alternating PD produced by either a **klystron** or a **magnetron**. The arrangement ensures that A, C, E . . . are positive when B, D, F . . . are negative and vice versa.

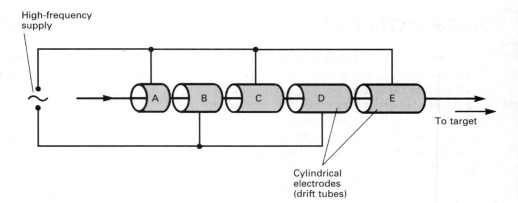

Fig. 8.7
Drift tube linear accelerator

A proton beam (from a Van de Graaff accelerator, say) is injected into the device and travels along the (common) axis of the cylinders. Any protons that reach the gap between A and B when A is positive and B is negative are accelerated across the gap and enter B with increased energy. All parts of B are at the same potential and therefore the protons travel through it at <u>constant</u> speed (hence 'drift tube'). The lengths of the cylinders are such that whenever the beam emerges into the gap between two cylinders, the cylinder ahead of it is negative and that which it has just left is positive. (Since the protons are accelerated at each gap, this requires that each cylinder is longer than the previous one.) It follows that the energy of the beam is increased every time the protons cross one of the gaps, and therefore a device with a large number of gaps can produce extremely high-energy beams using only moderate supply voltages. The Berkeley proton accelerator is 12 m long, has 47 drift tubes and accelerates protons to 31.5 MeV.

Linear accelerators have the following advantages:

(i)   They produce high-intensity, well-collimated beams.

(ii)   It is much easier to extract the beam than it is with a circular device such as a cyclotron.

(iii)   There is no magnet.

Travelling-Wave Accelerators

These accelerate charged particles by using the electric field component of a radio frequency electromagnetic wave travelling along a waveguide. The particles need to be moving at speeds close to that of light when they enter the device. They are therefore used to accelerate electrons, rather than protons, because these can easily be accelerated to suitable speeds using, for example, a Van de Graaff accelerator. Typically, electrons are injected at speeds of around $0.98c$. This leaves very little scope for any further increase in <u>speed</u> (since nothing can travel faster than light) and the increase in energy provided by the linear accelerator is due to the relativistic increase in <u>mass</u>.

The Stanford linear accelerator uses the travelling-wave principle and accelerates electrons (and positrons) to 50 GeV in a tube 2 miles long.

8.5 PROTON SYNCHROTRONS

The world's highest energy particle accelerators are **proton synchrotrons**. These devices overcome the problems associated with the relativistic increase in mass that limits the particle energies available from cyclotrons. Batches of protons (or antiprotons) accelerated by electric fields inside an evacuated tube travel round and round a circular path of large radius, guided by magnets placed at intervals around the tube. The strength of the magnetic field and the frequency of the electric field are gradually varied to compensate for the relativistic increase in mass.

The smaller circle indicates the tunnel housing the proton–antiproton collider at CERN. The larger circle is 27 km in circumference and marks the position of the Large Electron–Positron collider (LEP).

The large proton synchrotron at CERN (Geneva) has a diameter of 2.2 km and accelerates protons and antiprotons to 315 GeV. That at Fermilab in the USA provides protons and antiprotons with energies of 900 GeV.

8.6 FIXED-TARGET AND COLLIDING-BEAM EXPERIMENTS

If a particle collides with a stationary target, the total momentum of the particles emerging from the collision must be equal to that of the incident particle. It follows that these particles must have some kinetic energy, and therefore that some of the energy of the incident particle is not available to create new particles. This imposes a severe limitation on the energy available from these **fixed-target** experiments. For example, when a 500 GeV proton collides with a stationary proton, about 94% of the proton's kinetic energy goes into providing the kinetic energy of the collision products. For 1000 GeV protons it would be even worse – about 96% of the energy would be 'wasted'.

The problem does not arise with **colliding-beam** experiments. In these, beams of particles are caused to collide with their antiparticle counterparts moving in the opposite direction at the same speed. The total momentum of the colliding particles is therefore zero, and their total energy is available to create new particles.

A disadvantage of colliding-beam experiments is that collision rates are lower than in fixed-target experiments. The likelihood of collision is increased by using finely focused beams and by allowing the particles to accumulate in storage rings before the two beams finally meet.

CONSOLIDATION

An electric field changes the KE of a charged particle; a magnetic field cannot.

In an electric field

$$F = QE \qquad QV = \tfrac{1}{2} mv^2$$

In a magnetic field

$$F = BQv \qquad BQv = \frac{mv^2}{r} \qquad mv = BQr$$

Particle accelerators (linear and circular) use electric fields to accelerate charged particles.

Circular accelerators (e.g. cyclotrons and synchrotrons) use magnetic fields to direct charged particles along circular paths.

Cyclotrons accelerate protons. The maximum energy is limited to 25 MeV because the relativistic increase in mass of the protons causes them to become out of phase with the accelerating field. **Synchrotrons** overcome the problem of increasing mass and can accelerate protons and antiprotons to 900 GeV.

The Stanford linear accelerator accelerates electrons and positrons to 50 GeV.

In **colliding-beam** experiments the whole of the energy of the particles which collide is available to create new particles, but collision rates are lower than in **fixed-target** experiments.

QUESTIONS ON CHAPTER 8

1. A beam of electrons travelling with speed $1.2 \times 10^7\,\mathrm{m\,s^{-1}}$ in an evacuated tube is made to move in a circular path of radius 0.048 m by a uniform magnetic field of flux density $B = 1.4\,\mathrm{mT}$.
  (a) Calculate, in electronvolts, the kinetic energy of an electron in the beam. (The charge of an electron $= 1.6 \times 10^{-19}\,\mathrm{C}$ and $1\,\mathrm{eV} = 1.6 \times 10^{-19}\,\mathrm{J}$.)
  (b) A similar technique is used to accelerate protons to very high speeds. Protons with energies of 500 GeV can be held by magnetic fields in circular orbits of radius 2 km. Suggest why such a large radius orbit is necessary for high energy protons. [L]

2. (a) An electron (mass m, charge e) travels with speed v in a circle of radius r in a plane perpendicular to a uniform magnetic field of flux density B.
   (i) Write down an algebraic equation relating the centripetal and electromagnetic forces acting on the electron.
   (ii) Hence show that the time for one orbit of the electron is given by the expression $T = 2\pi m/Be$.
  (b) If the speed of the electron changed to $2v$, what effect, if any, would this change have on:
   (i) the orbital radius r,
   (ii) the orbital period T?

(c) Radio waves from outer space are used to obtain information about interstellar magnetic fields. These waves are produced by electrons moving in circular orbits. The radio wave frequency is the same as the electron orbital frequency.
(The mass of an electron is 9.1×10^{-31} kg, and its charge is -1.6×10^{-19} C.)
If waves of frequency 1.2 MHz are observed, calculate:
(i) the orbital period of the electrons;
(ii) the flux density of the magnetic field.
[O, '92]

3. The force required to keep a particle of mass m moving with speed v in a circular orbit of radius r is mv^2/r directed towards the centre of the circle. The magnitude of the force exerted on a particle carrying an electric charge Q and travelling at constant speed v, at right angles to a uniform magnetic field of flux density B, is BQv.
(a) Show that the charged particle will move in a circular path of radius mv/QB.
(b) Compare the paths of a neutron, a proton, and an alpha-particle when each is directed at right angles to the same magnetic field with identical velocities. [L, '91]

4. (a) What limits the energy to which protons can be accelerated using a cyclotron?
(b) Why are cyclotrons used to accelerate protons rather than electrons?
(c) A cyclotron used to accelerate protons has a radius of 40.0 cm, a magnetic field of 0.800 T and a dee voltage of 25.0 kV. Find **(i)** the frequency at which the cyclotron operates, **(ii)** the energy (in MeV) of the protons, **(iii)** the number of complete orbits made by the protons.
($e = 1.60 \times 10^{-19}$ C,
$m_p = 1.67 \times 10^{-27}$ kg.)

5. In a colliding-beam experiment a greater proportion of the colliding particles' total kinetic energy is available for producing new particles than in a fixed-target experiment. Explain why this is so. When might a fixed-target experiment be preferable? [L, '94]

6. A Van de Graaff accelerator is used to accelerate protons in a tube from rest to a velocity of $2.8 \times 10^7 \text{ m s}^{-1}$. Calculate
(a) the accelerating voltage,
(b) the power dissipation at the target, if the beam current is 1.0×10^{-4} A,

(c) the minimum length of the column, if the insulation breakdown field of the tube is $3.0 \times 10^6 \text{ V m}^{-1}$.
(Magnitude of charge on proton $= 1.60 \times 10^{-19}$ C, mass of proton $= 1.67 \times 10^{-27}$ kg.)
[J (specimen), '96]

7. This question is about the acceleration of protons to high energies in a proton synchrotron.
(a) Before injection into the synchrotron, the protons are accelerated from rest in an electrostatic accelerator through a total potential difference of 7.0 MV. Calculate their speed on injection into the synchrotron.
(b) In the synchrotron the protons move in a circular orbit of radius r, in the horizontal plane. The orbit is in a region of uniform vertical magnetic field B. Show that the orbital period of the protons is independent of the speed of the protons.
(c) According to the Theory of Relativity, at very high proton energies an increase in energy results in an increase in the mass of the protons. The speed of the protons is almost constant at nearly the speed of light. Suggest how a group of high-energy protons may be kept in an orbit of constant radius and periodic time as their energy is increased.
(Mass of proton $= 1.7 \times 10^{-27}$ kg,
$e = 1.6 \times 10^{-19}$ C.) [O & C, '93]

8. This question is about the acceleration of protons in the proton synchrotron at the European Centre for Nuclear Research (CERN) near Geneva.
You will find the data required on the next page and in the question.
Particle accelerators are used to increase the energy of charged particles such as protons and electrons. The accelerated particles are made to collide with other particles in a 'target' in order to investigate the structure of matter. Sub-atomic particles created in these collisions can be observed and studied.
One type of particle accelerator is the synchrotron. In this machine a magnetic field causes charged particles to travel in a circular path. The particles are accelerated by an electric field. As their momentum increases, the magnetic flux density is also increased to keep them travelling in a path of constant radius.
(a) Protons are injected into the 28 GeV proton synchrotron ring at CERN with an energy of 50 MeV (8×10^{-12} J).

Ignoring relativistic effects, show that:

(i)   the speed of a proton at injection is about $10^8 \, \text{m s}^{-1}$,

(ii) a proton takes about $6 \, \mu\text{s}$ to travel round the ring at this speed,

(iii) the momentum of a proton at injection is about $1.6 \times 10^{-19} \, \text{kg m s}^{-1}$.

(b) The accelerator ring is a vacuum pipe maintained at a very low pressure. It is 'filled' with protons by injecting a proton current of $100 \, \text{mA}$ for the $6 \, \mu\text{s}$ it takes for protons to make one revolution at the injection energy of $50 \, \text{MeV}$.

(i)   Calculate the number of protons injected.

(ii) Explain why the ring must be maintained at a very low pressure.

(c) Before the protons are accelerated, an electric field is used to group the protons in the ring into a number of bunches. The bunches of protons are then accelerated as they pass through each of 14 acceleration points spaced equally round the ring. An acceleration point is essentially a pair of electrodes between which an alternating voltage is applied.

(i)   Suggest why an alternating voltage is applied between the electrodes in order to accelerate the protons.

The proton bunches pass through the acceleration point when the voltage between its electrodes is about $4 \, \text{kV}$.

(ii) By how much does the energy of one proton increase in each revolution? (Give your answer in eV.)

The final energy of a proton is $28 \, \text{GeV}$.

(iii) Estimate the number of times a proton travels round the ring in acquiring this energy.

(iv) Explain briefly why linear accelerators (as opposed to ring accelerators) are not used to accelerate protons to these energies.

(d) (i) Show that the B-field required to maintain a proton of charge e in a circular path of radius r is proportional to the momentum p of the proton.

(ii) Estimate the B-field required to maintain $50 \, \text{MeV}$ protons within the CERN proton synchrotron.

(iii) Explain why the frequency of the accelerating voltage must be increased as the speed of the protons increases.

(e) When it reaches its maximum energy of $28 \, \text{GeV}$, the momentum of a proton is $1.6 \times 10^{-17} \, \text{kg m s}^{-1}$ and it is travelling at almost $3 \times 10^8 \, \text{m s}^{-1}$. The mass of any particle increases as its speed increases, although the effect only becomes important at speeds close to that of light.

Estimate:

(i)   by what factor the B-field must be increased during acceleration,

(ii) by what factor the mass of the proton increases during acceleration.

Protons from this accelerator can be injected into the Super Proton Synchrotron (SPS). This machine is $2.2 \, \text{km}$ in diameter and the ultimate energy of protons from it is $400 \, \text{GeV}\star$.

(iii) Explain why the B-field is increased but the frequency of the accelerating voltage is kept almost constant, as protons are accelerated in the SPS.

(Charge on proton $< 1.60 \times 10^{-19} \, \text{C}$, mass of proton $= 1.66 \times 10^{-27} \, \text{kg}$, $1 \, \text{MeV} = 1 \times 10^6 \, \text{eV}$, $1 \, \text{GeV} = 1 \times 10^9 \, \text{eV}$.)

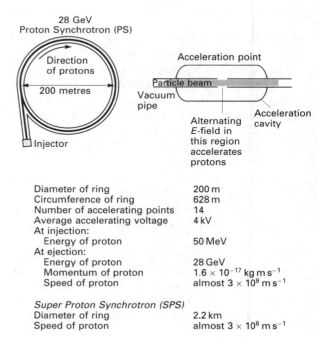

Diameter of ring	200 m
Circumference of ring	628 m
Number of accelerating points	14
Average accelerating voltage	4 kV
At injection:	
Energy of proton	50 MeV
At ejection:	
Energy of proton	28 GeV
Momentum of proton	$1.6 \times 10^{-17} \, \text{kg m s}^{-1}$
Speed of proton	almost $3 \times 10^8 \, \text{m s}^{-1}$
Super Proton Synchrotron (SPS)	
Diameter of ring	2.2 km
Speed of proton	almost $3 \times 10^8 \, \text{m s}^{-1}$

[O & C (Nuffield), '90]

*When protons and antiprotons are being accelerated at the same time, the particles have a maximum energy of $315 \, \text{GeV}$.

9
PARTICLE DETECTORS

9.1 THE GEIGER–MÜLLER TUBE (G–M TUBE)

A Geiger–Müller tube (Fig. 9.1) can be used to detect the presence of X-rays, γ-rays and β-particles. Tubes with very thin mica windows can also detect α-particles.

Fig. 9.1
A Geiger–Müller tube and circuit

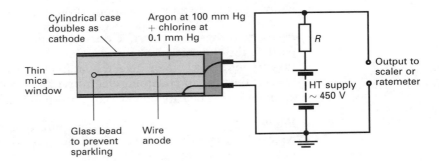

When one of these ionizing 'particles' enters the tube, either through the window or the wall, some of the argon atoms become ionized. The free electrons and positive ions which result are accelerated towards the anode and cathode respectively by the PD across the tube. The geometry of the tube is such that the electric field near the anode is very intense, and as electrons approach the anode they gain sufficient kinetic energy to produce further ionization. The electrons released by this 'secondary' ionization produce even more ionization so that there is soon a large number of electrons moving towards the anode – the resistance of the gas is said to have broken down. The positive ions are much more massive, and move much more slowly, than the electrons, and after about 10^{-6} s there are so many positive ions near the anode that the electric field around it is cancelled out. This prevents further ionization, and the **electron avalanche** and the associated anode current, cease to exist. Thus, the effect of a single ionizing 'particle' entering the tube is to produce a relatively large current pulse. The process is called **gas amplification** and as many as 10^8 electrons can be released as a result of a single ionizing event.

The positive ions move slowly towards the cathode. Some of the ions would release electrons from the cathode surface if they were allowed to collide with it. These electrons would initiate a second avalanche, and this would give rise to a third, and so on, so that a whole series of current pulses would be produced. This would make it impossible to know whether a second ionizing 'particle' had entered the tube. In order to ensure that only one pulse is produced by each 'particle' that enters it the tube contains a **quenching agent** – chlorine. (Bromine is used in some cases.)

The argon ions are neutralized as a result of collisions with chlorine molecules before they reach the cathode, and, in effect, their energy is used to dissociate the chlorine molecules rather than to release electrons from the cathode.

A resistor, R, of about 1 MΩ, is connected in series with the tube and the HT supply. The current pulse from the tube creates a voltage pulse of about 1 V across R, and this can be amplified and fed to a **scaler counter** or a **ratemeter**. A scaler registers the number of pulses it receives whilst it is switched on; a ratemeter indicates the rate at which it receives pulses and registers it in counts per second.

Immediately after a pulse has been registered there is a period of about 300 μs during which the tube is insensitive to the arrival of further ionizing 'particles'. This can be divided into two parts – the **dead time** and the **recovery time**. During the dead time the tube does not respond at all to the arrival of an ionizing 'particle'. The recovery time is the second stage of the period of insensitivity, and during this time pulses are produced but they are not large enough to be detected. The dead time is the time taken by the positive ions to move far enough away from the anode for the electric field there to return to a level which is large enough for an avalanche to start. The recovery time is the time which elapses while the argon ions are being neutralized by the quenching gas. The period of insensitivity limits the count rate to a maximum of about 1000 counts per second.

Apart from those which arrive during the period of insensitivity, almost every α-particle and β-particle that enters a Geiger–Müller tube is counted. γ-rays and X-rays are more likely to be detected indirectly, as a result of being absorbed by the walls of the tube and releasing electrons in the process, than by direct ionization of argon atoms. They are only weakly absorbed and only about 1% are detected.

Fig. 9.2 shows the way in which the count rate varies with the PD applied to a typical Geiger–Müller tube. When the applied PD is less than the **threshold voltage** there is not sufficient gas amplification to produce pulses which are large

Fig. 9.2
Variation of count rate with applied PD in a G–M tube

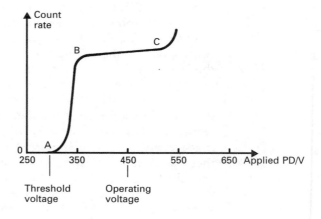

enough to be detected. Between A and B, the **proportional region**, the size of any particular pulse depends on the 'strength' of the initial ionization; some of the 'particles' which enter the tube produce less ionization than others and go undetected. In the **plateau region** (B to C) all the pulses have the same amplitude, irrespective of the 'strength' of the initial ionization. Every particle which produces any ionization at all is detected. This is the region in which the tube should be operated. If the voltage is increased beyond C, the quenching process becomes less and less effective and eventually a continuous discharge occurs.

9.2   THE IONIZATION CHAMBER

A common form of ionization chamber is shown in Fig. 9.3. It is essentially a metal can containing a small brass platform mounted on the upper end of a metal rod. The can forms one electrode of the device; the metal rod is the other. The simplest types contain air at atmospheric pressure and are capable of detecting the ionization produced by α-particles and <u>intense</u> sources of β-particles. The source of the radiation may be either outside the chamber, or inside it, in which case the platform provides a convenient means of supporting the source. When the source is outside the chamber a <u>gauze</u> 'lid' is used so that the particles can enter without appreciable absorption.

Fig. 9.3
Ionization chamber

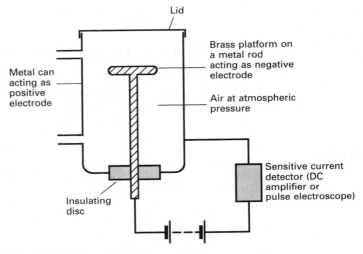

A typical α-particle produces 10^5–10^6 ion-pairs as it passes through the air in the chamber. The electrons are attracted to the can and the positive ions move to the central rod causing a current to flow in the external circuit. Fig. 9.4 illustrates the way in which this **ionization current** depends on the PD across the chamber. Between O and A the PD is not large enough to draw all the electrons and positive ions to their respective electrodes before some recombination has occurred. Between A and B the PD is large enough to prevent recombination but is not so high that it produces secondary ionization. The ionization current is said to have reached its **saturation value** (I_s). Beyond B the PD is large enough to cause secondary ionization. The PD at which an ionization chamber is operated should be such that the ionization current has its saturation value. Under such conditions:

(i)   the ionization current is independent of fluctuations in supply voltage, and

(ii) the ionization current is proportional to the rate at which ionization is being produced in the chamber. (The reader should contrast this with the case of the Geiger–Müller tube, where the output is proportional to the number of ionizing particles.)

Fig. 9.4
Variation of ionization current with PD across an ionization chamber

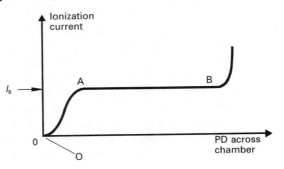

The ionization currents produced by the sources commonly used in schools are small – 10^{-9} A or less. This means that very sensitive current detectors are required; DC amplifiers and pulse electroscopes are suitable.

9.3   CLOUD CHAMBERS

These have not been used for many years in particle physics but are still of some interest. They are of two types.

The Diffusion Cloud Chamber

The cloud chamber (Fig. 9.5) displays the tracks of any ionizing agents which pass through it. It is superior to Wilson's earlier expansion cloud chamber (see p. 131) in that it does not have to be re-set before it can display a second track.

Fig. 9.5
Diffusion cloud chamber

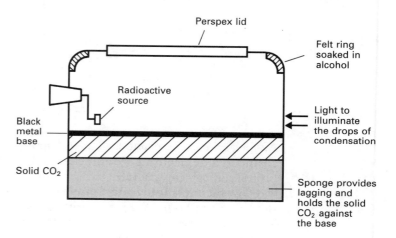

The base of the chamber is maintained at about $-80\,^\circ$C by the solid carbon dioxide there. The top of the chamber is at room temperature and so there is a temperature gradient between top and bottom. The air at the top of the chamber is saturated with alcohol vapour from the felt ring. The vapour continually diffuses downwards into the cooler regions so that the air there becomes supersaturated with alcohol vapour. The excess vapour in the supersaturated regions can condense only if there are nucleating sites present. Condensation occurs on ionized atoms in preference to neutral atoms, and so if an ionizing agent passes through the supersaturated air, the ions produced along its path act as nucleating sites. The path therefore shows up as a series of small drops of condensation.

α-particles leave dense, straight tracks; β-particles produce less ionization and give thinner tracks. The tracks of fast β-particles are straight; those travelling more slowly are easily deflected and leave tortuous tracks. γ-rays are uncharged and therefore must actually collide with an atom in order to ionize it. Since such collisions are rare, there is very little ionization along the path of the rays. However, when a collision does occur, the electron which is ejected has sufficient energy to ionize atoms along its path. All such paths originate on the path of the γ-rays and therefore, provided the beam is sufficiently intense, its presence can be detected (Fig. 9.6). X-rays produce a similar effect.

Fig. 9.6
The effect of γ-rays passing through a cloud chamber

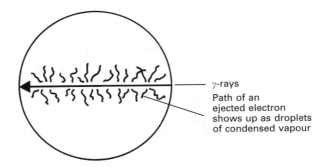

γ-rays

Path of an ejected electron shows up as droplets of condensed vapour

The Wilson Cloud Chamber

The Wilson cloud chamber, like the diffusion cloud chamber, shows up the paths of ionizing 'particles' which pass through it. Both types of chamber make use of the fact that a supersaturated vapour condenses more readily on ions than on neutral atoms. In the Wilson cloud chamber, a gas saturated with vapour is caused to undergo an adiabatic expansion; the gas cools and so becomes supersaturated with vapour. Various combinations of gas and vapour are used, e.g. air–water, air–alcohol, argon–water.

9.4 THE BUBBLE CHAMBER

The bubble chamber was invented by Glaser in 1951 and makes use of a superheated liquid. (A superheated liquid is one which is at a higher temperature than that at which it normally exists as a liquid under the prevailing pressure, and as such is unstable.) The ions produced in a superheated liquid by the passage of a <u>charged</u> particle act as nucleating sites on which bubbles form. The path of such a particle therefore shows up as a train of bubbles which can be photographed. The chamber is reset simply by increasing the pressure in it and collapsing the bubbles. **A neutral particle produces no ionization and therefore leaves no track**.

The Big European Bubble Chamber (BEBC) at CERN. Essentially a cylinder 3.7 m in diameter and containing liquid hydrogen at $-173\,°C$, it operated from 1973 to 1984 and was involved in 6.3 million photographs of particle interactions.

Bubble chambers are superior to cloud chambers in that the detecting medium is a liquid which, being more dense than the air (and other gases) used in cloud chambers, has a greater stopping power. This is particularly useful when particles with very high energies are being studied – such particles are likely to pass through a cloud chamber without producing any ionization at all.

One of the most commonly used liquids is hydrogen. This provides an opportunity to bombard protons (hydrogen nuclei) with high-energy particles and to study the particles which result from the collisions. By using magnetic fields to deflect the particles, and by making use of the laws of conservation of energy and momentum, the charges, masses and speeds of the particles can be deduced.

Bubble chambers were used extensively until the mid nineteen eighties when they began to be replaced by electronic detectors such as drift chambers. With these devices the data is recorded electronically, rather than photographically, and can be used to produce computer-generated images of the collision events.

9.5   THE SPARK COUNTER

This consists of a wire or mesh anode which is at a high PD relative to a nearby metal plate. If a charged particle passes between these electrodes, the air becomes conducting and a spark jumps across the gap revealing the presence of the particle.

9.6   THE SPARK CHAMBER

This is effectively a series of spark counters placed side by side, each of which produces a spark as an ionizing particle passes through it. The path of the particle therefore shows up as a trail of sparks.

Unwanted particle tracks are avoided by using an auxiliary detector to trigger the application of the necessary PD immediately after a particle of the type being studied has passed through the chamber.

9.7   INTERPRETATION OF BUBBLE CHAMBER PHOTOGRAPHS

A bubble chamber photograph shows the paths of <u>charged</u> particles. The paths are curved because bubble chambers are subjected to strong magnetic fields when they are in use. Interpreting the photographs is a complicated process but a certain amount of information can be gleaned simply by noting the points listed below.

(i)   It follows from equation [8.3] that the momentum of a particle of charge Q and mass m in a field of flux density B is related to the radius of curvature, r, of its track by

$$mv = BQr$$

Thus **the momentum of any given particle is proportional to the radius of its track**.

(ii)   It is usually possible to assume that $Q = \pm e$.

(iii) If the field is directed into the paper, negative particles curve clockwise; positive particles curve anticlockwise. (This follows from Fleming's left-hand rule.)

(iv) Neutral particles do not leave tracks because they produce very little ionization, but they travel in straight lines and therefore their paths can often be inferred from the presence of tell-tale gaps between the visible tracks. It is possible to determine the momentum of a neutral particle by applying the principle of conservation of momentum to the particles it has interacted with.

(v) Charge is conserved.

QUESTIONS ON CHAPTER 9

1. Describe the structure of a Geiger-Müller tube. Why are some tubes fitted with thin end windows? Why does the anode of a Geiger-Müller tube have to be made of a *thin* wire?
Explain the principle of operation of a cloud chamber. Describe and explain the differences between the tracks formed in such a chamber by alpha and beta particles.
A radioactive source has decayed to 1/128th of its initial activity after 50 days. What is its half-life?   [L]

2. The diagram below shows the track of a proton in a magnetic field. Copy this diagram and then draw and clearly label the tracks that would be produced by
(a) a proton of *higher* energy in the *same* magnetic field,
(b) a proton with the *same* energy in a *stronger* magnetic field.
Name *one* type of particle detector that might be used to record these tracks.

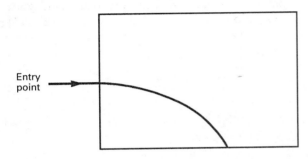

Explain why this detector cannot record the passage of neutrons.   [L (specimen), '92]

3. (a) The path of a charged particle through a cloud chamber appears as a thin white line.

What does this line consist of and how is it formed?
How may the tracks of beta particles be distinguished from those of alpha particles in the absence of any deflecting electric or magnetic fields?

(b) The diagram shows an exaggerated view of the track of a beta particle as it passes through a thin sheet of lead in a cloud chamber. A uniform magnetic field acts *into* the paper, through the whole region shown in the diagram, and the beta particle moves in the plane of the paper.

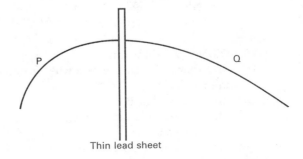

Thin lead sheet

(i) Is the beta particle travelling in the direction P to Q or Q to P? Explain your reasoning.

(ii) Does this beta particle carry a positive or negative charge? Explain your reasoning.

(c) A $^{40}_{19}$K atom has mass 39.964 001 u and a $^{40}_{20}$Ca atom has mass 39.962 582 u where 1 u equals 1.6604×10^{-27} kg. The speed of light in vacuum is $3.00 \times 10^8 \, \mathrm{m \, s^{-1}}$.

(i) Write down an equation which describes the decay of a $^{40}_{19}$K atom into a $^{40}_{20}$Ca atom and identify the particle emitted by the nucleus.

(ii) Why, in the decay process, does the number of extranuclear electrons increase by one?

(iii) Calculate the energy released from the $^{40}_{19}K$ nucleus during the decay.

(iv) Describe one possible mode of decay of a $^{40}_{19}K$ nucleus into a $^{40}_{18}Ar$ nucleus.

[L]

4. (a) A G–M tube is exposed to a constant flux of alpha particles. The graph below shows how the recorded count rate depends on the potential difference across the tube.

Draw and label a diagram of a G–M tube. Outline its working principle with reference to what happens when an alpha particle enters the tube. Explain why there is an upper limit to the rate at which a G–M tube can detect α-particles.

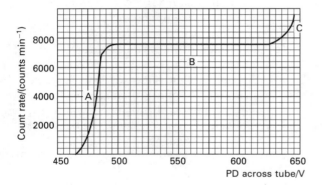

How do you account for:

(i) the sharp rise in the recorded count rate at A,

(ii) the 'plateau' at B, and

(iii) the uncontrolled rise in the recorded count rate at C?

State what potential difference e you would choose for the Geiger counter whose response is shown in the graph. Give one good reason for your choice.

(b) A small amount of ^{24}Na is smeared on to a card and its activity falls by 87.5% in 45 h. What is the half-life of ^{24}Na?

Describe how you would use a G–M tube in conjunction with a suitable counter to measuure the half-life of ^{24}Na. Explain carefully how the result is found from the measurements.

$$\left(\text{Decay constant} = \frac{0.693}{\text{half-life}}. \right) \qquad [L]$$

5. The diagram shows the track of a charged particle in a magnetic field. The field is at right angles to the plane of the paper, and its direction is out of the plane of the paper. AB is a thin sheet of lead through which the particle passes.

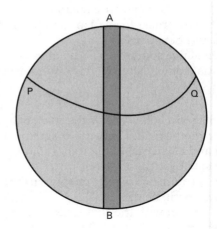

Deduce the direction of movement of the particle and the sign of the charge on the particle. Explain clearly how you made your deductions. [L, '94]

VALUES OF SELECTED PHYSICAL CONSTANTS

Quantity	Symbol	Value
Speed of light in vacuum	c	$2.99792 \times 10^8 \, \text{m s}^{-1}$
Planck's constant	h	$6.62608 \times 10^{-34} \, \text{J s}$
Electronic charge	e	$1.60218 \times 10^{-19} \, \text{C}$
Mass of electron	m_e	$9.10939 \times 10^{-31} \, \text{kg}$
		$5.48580 \times 10^{-4} \, \text{u}$
		$5.10999 \times 10^{-1} \, \text{MeV}/c^2$
Mass of proton	m_p	$1.67262 \times 10^{-27} \, \text{kg}$
		$1.00728 \, \text{u}$
		$9.38272 \times 10^2 \, \text{MeV}/c^2$
Mass of neutron	m_n	$1.67493 \times 10^{-27} \, \text{kg}$
		$1.00867 \, \text{u}$
		$9.39566 \times 10^2 \, \text{MeV}/c^2$
Unified atomic mass unit	u	$1.66054 \times 10^{-27} \, \text{kg}$
		$9.31494 \times 10^2 \, \text{MeV}/c^2$
Avogadro constant	N_A	$6.02214 \times 10^{23} \, \text{mol}^{-1}$
Permittivity of vacuum	ε_0	$8.85419 \times 10^{-12} \, \text{F m}^{-1}$
	$1/4\pi\varepsilon_0$	$8.98755 \times 10^9 \, \text{m F}^{-1}$
Boltzmann's constant	k	$1.38066 \times 10^{-23} \, \text{J K}^{-1}$

ANSWERS TO END OF CHAPTER QUESTIONS

The Examination Boards accept no responsibility whatsoever for the accuracy or method of working in the answers given. These are the sole responsibility of the author.

CHAPTER 1

3. 6.0 MeV
4. (c) $1.0 \times 10^{-6} \, \text{s}^{-1}$ (d) 1.0×10^{18}
5. $8.8 \times 10^8 \, \text{Bq}$
6. $50 \, \text{s}^{-1}$
7. (b) (i) 6.3×10^8 (ii) $2.0 \times 10^{-6} \, \text{s}^{-1}$ (iii) $1.3 \times 10^3 \, \text{Bq}$
8. $11 \, \mu\text{g}$
9. 4.3 hours
10. $2.8 \times 10^{11} \, \text{Bq}$
11. (a) 6.0×10^{11} (b) 6.5×10^{11}
12. 140 days; $a = 210$, $b = 84$, $c = 4$, $d = 2$, $c = 0$, $f = 0$
13. 6.0 m
14. (c) (i) $6.2 \times 10^{11} \, \text{s}$
15. $> 57 \, \mu\text{Ci}$
16. (d) $\frac{1}{32}$
17. 2.0 m; 24 hours
18. (b) $4.6 \times 10^{-4} \, \mu\text{g}$
19. 5.52×10^5
21. (a) (i) 5.0×10^{18} (ii) $8.8 \times 10^{-13} \, \text{s}^{-1}$
    (b) (i) $\frac{1}{4}$ (ii) 50 000 years

CHAPTER 2

1. (c) 1.1×10^{11} years
3. (b) $1.824 \, \text{u}$
4. (b) (i) 7.57 MeV (ii) 7.88 MeV
5. (b) 5.07 MeV, 1.20 MeV
7. (a) 28.3 MeV (b) 23.8 MeV
8. 186 MeV
9. $2.79 \times 10^{-12} \, \text{J}$
11. (a) $3.0 \times 10^6 \, \text{s}^{-1}$, $1.3 \times 10^{-9} \, \text{g}$
12. $2.6 \times 10^{-30} \, \text{kg}$, $2.3 \times 10^{-13} \, \text{J}$

CHAPTER 3

1. (a) $2.7 \times 10^{-14} \, \text{m}$
2. (b) 24 MeV
3. (a) $6.8 \times 10^{-15} \, \text{m}$ (b) $3.7 \times 10^{-11} \, \text{J}$
6. (c) $2.32 \times 10^{-15} \, \text{m}$, $2.56 \times 10^{-15} \, \text{m}$
7. (b) (i) $2.24 \times 10^{-19} \, \text{kg m s}^{-1}$ (ii) $2.96 \times 10^{-15} \, \text{m}$
    (iii) $2.60 \times 10^{-15} \, \text{m}$
    (c) (i) $3.9 \times 10^{-15} \, \text{m}$ (ii) $1.1 \times 10^{-42} \, \text{m}^3$

CHAPTER 4

2. (b) (ii) 2.62 MeV (iv) $1.22 \times 10^{-12} \, \text{m}$ (v) 2.61 MeV
3. (b) (ii) 0.81 MeV
4. (c) (i) $2.5 \times 10^{-14} \, \text{J}$ (ii) $9.2 \times 10^{-5} \, \text{kg}$
5. (a) (ii) $6.9 \times 10^{-12} \, \text{m}$
9. (a) (i) 12 MeV (iii) 28.4 MeV

CHAPTER 5

4. (a) 7.6 MeV
8. (c) (i) 9.4×10^{19} (ii) $2.9 \times 10^{-3} \, \text{kg}$
12. (b) 17.6 MeV (c) 3.5 MeV and 14.1 MeV
13. (b) (i) $1.5 \times 10^{27} \, \text{MeV}$ (ii) 2.9 kg
15. (d) (i) $5.2 \times 10^{-13} \, \text{J}$ (ii) 3.3 Mev
16. (b) $7.01 \times 10^{13} \, \text{J}$
17. (b) $4.30 \times 10^{-12} \, \text{J}$

CHAPTER 6

1. (b) (i) $1.6 \times 10^{-13} \, \text{J}$ (ii) $1.2 \times 10^{-12} \, \text{m}$
5. (a) 23 mN
7. (a) (i) 88
9. (a) π^+ (b) K^0 (c) $\bar{\nu}_\mu$ (d) K^+
11. (c) $205 \, \text{MeV}/c$
16. $3.0 \times 10^{-10} \, \text{J}$, $1.3 \times 10^{-15} \, \text{m}$

CHAPTER 7

2. (a) $3.1 \times 10^{25} \, \text{m}$ (b) 3.3×10^9 light-years

CHAPTER 8

1. (a) $4.0 \times 10^2 \, \text{eV}$
2. (c) (i) $8.3 \times 10^{-7} \, \text{s}$ (ii) $4.3 \times 10^{-5} \, \text{T}$
4. (c) (i) 12.2 MHz (ii) 4.91 MeV (iii) 98
6. (a) 4.1 MV (b) $4.1 \times 10^2 \, \text{W}$ (c) 1.4 m
7. (a) $3.6 \times 10^7 \, \text{m s}^{-1}$
8. (b) (i) 3.75×10^{12} (c) (ii) 56 keV (iii) 5.0×10^5
    (d) (ii) $1.0 \times 10^{-2} \, \text{T}$ (e) (i) $100\times$ (ii) $33\times$

CHAPTER 9

1. 7.1 days
3. (c) (iii) $2.1 \times 10^{-13} \, \text{J}$
4. (b) 15 hours

ANSWERS TO QUESTIONS 1A–8B

QUESTIONS 1A

1. **(a)** 26 **(b)** 26 **(c)** 30 **(d)** 56
   (e) 26 **(f)** 26 **(g)** 30
2. **(a)** 63, 65 **(b)** 62.9296 u, 64.9278 u **(c)** 63.5490

QUESTIONS 1B

1. **(a)** α, β **(b)** γ **(c) (i)** α **(ii)** α, β **(d)** β **(e)** α
   (f) α **(g)** γ **(h)** β **(i)** α **(j)** γ **(k)** β **(l)** β
   (m) α, β **(n)** β
2. $a = 4$, $b = 2$, $c = 220$, $d = 86$, $e = 210$, $f = 0$, $g = 84$, $h = 0$, $i = 0$

QUESTIONS 1C

1. $\frac{7}{8}$
2. $240\,\mathrm{s}^{-1}$
3. $1.2 \times 10^{12}\,\mathrm{Bq}$
4. 186 years
5. **(a)** 1.4×10^{20} **(b)** $7.9 \times 10^{16}\,\mathrm{Bq}$ **(c)** $9.8 \times 10^{15}\,\mathrm{Bq}$
   (d) 1:7
6. $1.3 \times 10^{-12}\,\mathrm{g}$

QUESTIONS 2A

1. **(a)** $1000\,\mathrm{eV}$ **(b)** $1.602 \times 10^{-16}\,\mathrm{J}$ **(c)** $1.875 \times 10^7\,\mathrm{m\,s}^{-1}$
2. $8.4 \times 10^6\,\mathrm{m\,s}^{-1}$
3. $300\,\mathrm{eV}$
4. $30\,\mathrm{V}$
5. $1.4 \times 10^7\,\mathrm{m\,s}^{-1}$

QUESTIONS 2B

1. **(a)** $1.93542\,\mathrm{u}$ **(b)** $7.57\,\mathrm{MeV}$

QUESTIONS 2C

1. **(a)** $0.00622\,\mathrm{u}$ **(b)** $5.79\,\mathrm{MeV}$
2. **(a)** $0.00129\,\mathrm{u}$ **(b)** $1.20\,\mathrm{MeV}$
3. **(a)** 51 **(b)** $5.11\,\mathrm{MeV}$

QUESTIONS 2D

1. $5.02\,\mathrm{MeV}$
2. $1.50 \times 10^{27}\,\mathrm{MeV}$

QUESTIONS 3A

1. **(a)** $8.1\,\mathrm{fm}$ **(b)** $2.2\,\mathrm{fm}$
   (c) (i) $3.5 \times 10^{-12}\,\mathrm{J}$ **(ii)** $22\,\mathrm{MeV}$

QUESTIONS 6A

1. **(a)** $5.0\,\mathrm{MeV}$ **(b)** $13.0\,\mathrm{MeV}$ **(c)** $8.0\,\mathrm{MeV}$
   (d) $12c/13 \,(= 0.92c)$ **(e)** $13.0\,\mathrm{MeV}/c^2$
2. $41\,\mathrm{MeV}$
3. **(a)** $62.5\,\mathrm{MeV}$ **(b)** $62.5\,\mathrm{MeV}/c$
4. **(a)** $1135\,\mathrm{MeV}\,(\Lambda^0)$, $537\,\mathrm{MeV}\,(\mathrm{K}^-)$
   (b) $19.5\,\mathrm{MeV}\,(\Lambda^0)$, $42.5\,\mathrm{MeV}\,(\mathrm{K}^-)$

QUESTIONS 6B

1. **(a)** $2.83 \times 10^{20}\,\mathrm{Hz}$ **(b)** $8.74 \times 10^{20}\,\mathrm{Hz}$
2. **(b)** $1.57 \times 10^{20}\,\mathrm{Hz}$
3. $0.90\,c$

QUESTIONS 6C

1. The reactions that cannot occur are: (a), (c), (d), (e), (f), (i), (k) and (p)
2. **(a)** Ω^- **(b)** K^+ **(c)** n

QUESTIONS 8A

1. $2.77 \times 10^5\,\mathrm{m\,s}^{-1}$
2. $8.5\,\mathrm{cm}$

QUESTIONS 8B

1. **(a)** $27\,\mathrm{MeV}$ **(b)** $13.5\,\mathrm{MeV}$ **(c)** $27\,\mathrm{MeV}$
2. 50
3. **(a)** $1.05\,\mathrm{T}$ **(b)** $8.44\,\mathrm{MeV}$
4. $3.2\,\mu\mathrm{s}$

INDEX

Parentheses indicate a page where there is a minor reference.